**Library of Congress Cataloging-in-Publication Data**

Liquid membranes: chemical applications/editors, Takeo Araki and
   Hiroshi Tsukube.
       p.   cm.
   Includes bibliographical references.
   ISBN 0-8493-5314-9
   1. Liquid membranes.   I. Araki, Takeo.   II. Tsukube, Hiroshi,
1953-
QD562.I63L56   1990                                                          89-25388
541.3'723--dc20                                                              CIP

   Direct all inquiries to CRC Press, Inc., 2000 Corporate Blvd., N.W., Boca Raton, Florida, 33431.

© 1990 by CRC Press, Inc.

International Standard Book Number 0-8493-5314-9

Library of Congress Card Number 89-25388
Printed in the United States

# PREFACE

Advances in the study of liquid membranes have been very rapid in recent years, and investigations have been made in a variety of disciplines on such aspects as the molecular design of synthetic carriers, mechanisms of mass transfer processes, design of modules for the technological uses of separation of gas and organic compounds, desalination, enrichment of resources, drug designs, modeling of the biological cell membranes, and medicinal analyses.

Liquids, representing a peculiar compromise between order and disorder, occur in a variety of states. The cohesive forces in liquids are sufficiently strong to give a condensed state, but not so strong that they prevent the translational motion of individual molecules. Thermal motions cause a disorder in liquid but without complete loss of the regularity of its structure, depending on the nature of the compound forming the liquid. These fundamentally unique characteristics of liquids will greatly broaden the application of liquid membranes in sophisticated devices in modern technology.

In organizing this volume, the editors have recognized that readers are interested in becoming acquainted with currently important subjects, especially the chemical applications of liquid membranes. Emphasis has been placed on molecular design of the carriers and chemical functionalization of synthetic liquid membranes. Some chapters detail items of biomimetic and practical interest in membrane transport phenomena. Although not all liquid membrane applications are covered, newer aspects and recent developments in the molecular functionalizations of these membranes and their special advantages in chemical processes are presented.

The book is primarily aimed at researchers and students in organic, physical, analytical, and inorganic chemistry, and those in certain fields of engineering chemistry. We hope, however, that it will also be of value to scientists in the areas of biology, biochemistry, physiology, and bioengineering. Extensive literature references will be found in each chapter, and will provide a literature guide or progress report helpful for those who are now planning to enter this field.

We would like to express our thanks to the contributors and the management of CRC Press, Inc. for making publication possible. We are grateful to Mrs. Jane Clarkin for her kind reviewing of several manuscripts. Our thanks are also given to Mr. Hisanori Ando and Mr. Hideaki Asai, students of Shimane University, for their help in many stages of the preparation.

**Takeo Araki**
**Hiroshi Tsukube**

# THE EDITORS

**Takeo Araki, Ph.D.,** is Professor of Polymer Chemistry at the Kyoto Institute of Technology.

Dr. Araki obtained his training at the Kyoto University, Kyoto, Japan, receiving the B.S. degree in 1957, and the Ph.D. degree in 1963 from Kyoto University. He served as Assistant Professor, Lecturer, and Associate Professor in the Department of Macromolecular Science at Osaka University (Osaka, Japan) from 1962 to 1980, and as Professor in the Department of Chemistry at Shimane University (Matsue, Japan) from 1980 to 1989. From 1984 to 1988 he served as a Member of Councilmen of the Shimane University. He was a Post-doctoral fellow at the University of Manchester (Manchester, U.K.) from 1967 to 1968. It was in 1989 that he assumed his present position.

Dr. Araki is a member of the American Chemical Society, in the Divisions of Polymer Science, Physical Chemistry, and Biochemistry; the Chemical Society of Japan; the Society of Polymer Science, Japan; the Society of Synthetic Organic Chemistry, Japan; the Japan Society of Energy and Resources; and the Kinki Chemical Society. He has served or is serving as an executive member of the Chemical Society of Japan; the Society of Polymer Science, Japan; and the Society of Synthetic Organic Chemistry, Japan. He served as a President of CSJ Annual Meeting of Chugoku-Shikoku Division in 1987, and on the Editorial Board of Kagaku To Kogyo (Chemistry and Chemical Industry) from 1982 to 1984, and has served on the Editorial Board of Chemistry Letters since 1989. He has been a recipient of many research grants from the Ministry of Education, Science, and Culture.

Dr. Araki is the author of approximately 100 papers and has been the author or co-author of eleven books. His current major research interests relate to specificity in molecular interactions and its applications, including chemistry of functionalized oligomeric compounds, synthesis of new functional polymers, and the relation between centrifugal countercurrent separation and liquid membrane transport systems.

**Hiroshi Tsukube, Ph.D.,** is associate professor of Chemistry at the College of Liberal Arts and Science, Okayama University, Japan.

Dr. Tsukube was born in Osaka in 1953 and attended Osaka University, graduating with a B.S. in polymer chemistry in 1975. He then attended Kyoto University, worked with professor Kazuhiro Maruyama, and gained a Ph.D. in organic chemistry. In 1981, he was appointed lecturer at Okayama University and was promoted to associate professor in 1984.

Dr. Tsukube is a member of The Chemical Society of Japan; The Society of Synthetic Organic Chemistry, Japan; and The Society of Polymer Science, Japan. He received the Nodzu Memorial Award in 1982 and the Progress Award from the Chemical Society of Japan in 1987.

Dr. Tsukube is the author of more than 60 papers and has been the co-author of five books. His research interests include synthesis of new host molecules, design of carrier-mediated membrane transport, and functional modification of natural ionophores.

# CONTRIBUTORS

**Takeo Araki, Ph.D.**
Professor
Polymer Science and Engineering
  Department
Kyoto Institute of Technology
Kyoto, Japan

**Jerald S. Bradshaw, Ph.D.**
Professor
Department of Chemistry
Brigham Young University
Provo, Utah

**Ronald L. Bruening, Ph.D.**
Vice-President
Department of Research and Development
IBC Advanced Technologies
Orem, Utah

**Kazuhisa Hiratani, Ph.D.**
Senior Researcher
Industrial Products Research Institute
Agency of Industrial Science and
  Technology
MITI
Tsukuba, Ibaraki, Japan

**Yoshihisa Inoue, Ph.D.**
Associate Professor
Basic Research Laboratory
Himeji Institute of Technology
Himeji, Hyogo, Japan

**Reed M. Izatt, Ph.D.**
Professor
Department of Chemistry
Brigham Young University
Provo, Utah

**Naoki Kamo, Ph.D.**
Professor
Department of Pharmaceutical Sciences
Hokkaido University
Sapporo, Japan

**John D. Lamb, Ph.D.**
Associate Professor
Department of Chemistry
Brigham Young University
Provo, Utah

**Tsutomu Matsuda, Ph.D.**
Professor
Department of Organic Chemistry
Kyushu University
Fukuoka, Japan

**Yoshio Okahata, Ph.D.**
Associate Professor
Department of Polymer Chemistry
Tokyo Institute of Technology
Ookayama, Meguro-ku, Tokyo, Japan

**Seiji Shinkai, Ph.D.**
Professor
Department of Organic Synthesis
Kyushu University
Fukuoka, Japan

**Hiroshi Tsukube, Ph.D.**
Associate Professor
Department of Chemistry
College of Liberal Arts and Science
Okayama University
Okayama, Japan

**Tomohiko Yamaguchi, Ph.D.**
Researcher
National Chemical Laboratory for
  Industry
Agency of Industrial Science and
  Technology
MITI
Tsukuba, Ibaraki, Japan

**Kenichi Yoshikawa, Ph.D.**
Associate Professor
College of General Education
Nagoya University
Nagoya, Japan

# TABLE OF CONTENTS

Chapter 1

# INTRODUCTION

**Takeo Araki**

# TABLE OF CONTENTS

# I. DEFINITION OF LIQUID MEMBRANES AS USED IN THIS VOLUME

People encounter liquid membranes or films in various forms in daily life. An oil layer on the water surface is a typical organic liquid membrane of an immiscible liquid phase. Bubbles of beer and the foam of soap, detergent, or surfactant solutions are familiar liquid films separating two gaseous phases. Oil films coated on a metal surface have been popularly used in rust protection, lubrication, or to prevent adhesion. Liquid film can be seen on many types of adhesive tape. Oil penetration has been one method of creating water-repellent finish for paper, textiles, wood, and clay. One can find water membranes absorbed by a damp cloth and also adsorbed in porous silica gel, both of which involve a chemical bond between a liquid and a solid. Detergents and surfactants dissolved in water form a regularly arranged aggregation of molecules called micelles, a type of structured liquid membrane with charge. These versatile occurrences of rather familiar liquid films and membranes have broad industrial and everyday applications, but these rather classical applications are quite different from modern liquid membrane processes.

Current students know that many types of biological cell membranes consist of regularly arranged double lipid layers. Since the lipid is a naturally occurring surfactant, the cell membranes are also a type of liquid membrane.

It is impossible to draw an exact distinction between liquid films and liquid membranes. In this volume, however, the term "liquid membranes" is limited to its use in contemporary scientific and technological journals. Liquid membranes here are media consisting of liquid films or membranes through which selective mass transfers of gases, ions, and molecules occur via permeation and transport processes, thus rendering the membranes functional. The term is further used for systems that have this potential in future technology.

A special note should be made here discriminating between "emulsion" and "emulsion liquid membrane" (or "liquid surfactant membrane"), because appearance of emulsion liquid membrane systems is almost identical with emulsion. A microglobule in emulsion is a tiny liquid particle covered by a surfactant film as the structured liquid. Such simple emulsions are now rarely included among liquid membranes. In a special case, however, a microglobule in the emulsion can be coated by an additional thin liquid film. For example, when an oil-in-water (O/W) emulsion is introduced into an oily medium by stirring, a multiple emulsion is founded with the sequence of phases oil-in-water-in-oil (O/W/O). The innermost oily phase is separated from the continuous oily phase by a skin of aqueous liquid membrane. Similarly, a W/O emulsion forms a W/O/W multiple emulsion particle coated by a skin of oily liquid membrane. Mass transfer by permeating through the new skin and the surfactant layers from the continuous outermost phase into the innermost phase (or vice versa) frequently differs from that through simple emulsion. Hence, it becomes a very useful device for chemical technology and is now a very important type of liquid membrane.[1]

The most central common feature in a variety of liquid membrane systems is the speed of the mass transfer processes involved compared with those through the solid membrane systems. These mass transfer processes are largely governed by permeation, dissolution, or diffusion phenomena, and only slight external energy (mainly for rotation, stirring, pumping, etc.) is needed. The rapidity and cost-saving aspects of the liquid membrane processes have attracted researchers of modern science and technology. The use of a mobile carrier, a transport catalyst, or a mediator which dissolves only in liquid membrane further increases the efficiency due to its high selectivity associated with the reversible specific binding or complexing process involved. Facilitated (or back-pack) transport is thus opened, and since then, the search for effective carriers has been one of the central interests in modern liquid membrane chemistry and technology.

## II. "LIQUID MEMBRANE TECHNOLOGY"

A new epoch in the application of liquid membrane began in the 1960s when Li et al. patented the use of emulsion liquid membrane systems for industrial scale desalination and the separation of hydrocarbons.[1] Based on a difference in the rate of mass transfer through liquid membranes, the transfer and selective efficiencies were shown by these researchers to be remarkably higher than those through a solid polymer membrane. The liquid membrane process was originally described in a book published in 1972.[2]

The first application of a supported liquid membrane system in modern analytical separation technique was probably in the 1950s in gas-liquid partition chromatography (gas-liquid chromatography),[3] in which a nonvolatile liquid was coated on a porous Celite support in the stationary phase. The concept of reaction gas chromatography was developed in this field, in which a catalyst in the liquid phase facilitated separation.

However, the work using the supported liquid membrane (or immobilized liquid membrane) reported by Ward and Robb in 1967[4] provided new potential for industrial separation. They showed that an aqueous bicarbonate-carbonate solution membrane supported in porous cellulose acetate film was 4100 times more permeable by $CO_2$ than by $O_2$. They also found that the transport of $CO_2$ could be increased by addition of a catalyst (sodium arsenite) to the liquid membrane for the hydrolysis of $CO_2$, concentrating the $CO_2$ present in the atmosphere of a manned spacecraft at approximately 1 to over 95% in only one passage through the membrane. This finding was also very important in recognition of the potential of reactive or responsive liquid membranes.

Neutron bombardment followed by etching can yield cylindrical pores (porosity: 40 to 80%) with a narrow distribution of diameter (0.01 to $10^{-6}$ m) in polymer films down to $10^{-4}$ m in thickness. Advances in hollow fiber techniques and plasma polymerization processes have greatly contributed to liquid membrane technology, so that now a variety of sophisticated liquid membranes including ultrathin liquid films supported on a desired synthetic polymer can be obtained almost at will.

The scientific publication *Chemical Abstracts* covered more than 100 articles under the category "liquid membrane" in each of 1987 and 1988, most of which were relevant to theory and applications in so-called liquid membrane technology for separation of metals, gases, and organic compounds. Advances in this field are actually of great importance,[5] and a plant for desalination using an emulsion liquid membrane system was very recently constructed in Australia as a culmination of cooperative international research.

A variety of problems remain to be solved, however. Among them, development of the type of module is a prime factor in assuring the possible operating stability of the liquid membrane used. New theoretical concepts are required to resolve problems ocurring with the new module, and those are highly empirical. Recognition of this has been well documented in review articles published in every industrial country, and for this reason details of liquid membrane technology are largely omitted in this volume. An overview of this subject was published in 1987.[6] Simultaneous removal and concentration of dissolved species from a diluted solution, a process called creeping film pertraction,[7] has recently been developed using a three-phase liquid membrane.

## III. NEWER APPLICATIONS OF LIQUID MEMBRANES

A variety of carriers involving proteins, special antibiotics, and many other natural compounds present in the cell membranes can select a desired substrate via any type of molecular recognition (specific binding), transport it through the cell membrane (rapid transfer), and then release it at the interfaces of the cell wall. These carriers are maintained almost indefinitely in the cell membranes, and their movement is induced by dynamic liquid-like thermal motions of the hydrophobic moieties in lipid molecules. No significant energy supply is required here, and a very small amount of a carrier is sufficient for recycling transport thousands of times. The carrier frequently produces electrical or chemical responses upon binding which can be detected by a neighboring biological molecular system with subsequent propagation to other tissues. The role of biological carrier is thus a key factor of tremendous importance in life functions.

In the development of the artificial carriers for use as chemical devices in future technology, a knowledge of cell membranes is necessary, and the behavior of the biological carrier compound acting as sensor or receptor should be modeled. Simpler ideas of the immobilization of enzymes[8] or blood[9] in liquid membranes were developed earlier and are now under extensive study for their applicability in areas of biotechnology and for medical uses such as anticancer drugs.

The first important indication of mimicking the biological specific ion recognition in an artificial liquid membrane system was reported by Cussler in 1971.[10] He was successful in selective transport of alkali metal ions dissolved in aqueous phase through a chloroform liquid membrane containing crown ether as carrier. Macrocyclic host-guest chemistry first introduced by Pedersen,[11] Lehn and his colleagues,[12] and Cram's group[13] has been greatly developed during the past two decades. Aiming at new functions and utilization, macrocyclic molecular designs have progressed extensively with synthetic innovations. Acyclic host-guest chemistry has also advanced, but to a lesser extent. Host-guest chemistry has been utilized in liquid membrane systems in place of newer chemical applications including biomimetic systems.

This publication concentrates largely on the last subject, starting with a chapter by an author who has strong interests in cell membranes. The chapters that follow are collective reviews prepared by researchers now actively manipulating various modern chemical sensing devices using host-guest and biomimetic chemistry. Special emphasis is given to the design of carrier compounds. Molecular design of surfactants in bilayer liquid membranes is also a topic. The final section explains an approach to a chromatographic process (CPC and CCC) leading to industrial scale isolation/purification/enrichment, a new type of multistage liquid membrane transport system.

## IV. CONCLUSION

While there is much of interest for the reader to digest, these pages offer only a fraction of the ongoing research now being conducted in this vital field. For instance, a highly important medical application of liquid membrane techniques to studies on skin penetration modeling[14] is not included. The reader will, however, recognize the breadth and depth of the areas in which liquid membrane systems or processes are now being implemented, and will, I feel sure, be stimulated by what he learns here.

## ACKNOWLEDGMENTS

The author appreciates those at the Ministry of Education, Science, and Culture who supported financially my research on the relation between liquid membrane transport and centrifugal partition chromatography by the Grant-in-Aid for General Scientific Research (No. 61470026). The review of this research will be made in Chapter 7.5.

# REFERENCES

1. **Li, N.N.,** U.S. Patent 3,454,489, 1966. **Li, N.N. and Somerset, N.J.,** Separating hydrocarbons with liquid membranes, U.S. Patent 3,410,794, 1968. **Li, N.N.,** Permeation through liquid surfactant membranes, *AICHE J.,* 17, 459, 1971.
2. **Li, N.N. and Shrier, A.L.,** *Recent Developments in Separation Science,* Vol. 1, Li, N.N., Ed., CRC Press, Cleveland, 1972, 163.
3. **James, A.T. and Martin, A.J.P.,** The separation and micro-estimation of volatile fatty acids from formic acids to dodecanoic acids, *Analyst,* 77, 915, 1952.
4. **Ward, W.J., III and Robb, W.L.,** Carbon dioxide-oxygen separation: Facilitated transport of carbon dioxide across a liquid film, *Science,* 156, 1481, 1967.
5. **Marr, R. and Kopp, A.,** Liquid membrane technology — a survey of phenomena, mechanisms, and models, *Int. Chem. Eng.,* 22, 44, 1982.
6. **Noble, R.D. and Way, J.D.,** Eds., *Liquid Membranes: Theory and Applications,* ACS Symposium Ser., No. 347, Am. Chem. Soc., Washington, D.C., 1987.
7. **Boyadzhiev, L. and Lazarova, Z.,** Creeping film pertraction — a new liquid membrane technique, *Izv. Khim. Inst. Bulg. Akad. Nauk.,* 20, 359, 1987; *Chem. Abst.,* 108: 63202m, 1988.
8. **Shrier, A.L.,** Liquid membranes in enzyme research, *Biotechnol. Bioeng. Symp.,* 3, 323, 1972. **May, S.W. and Li, N.N.,** *Biochem. Biophys. Res. Commun.,* 47, 1179, 1972.
9. **Li, N.N. and Asher, W.J.,** Blood oxygenation by liquid membrane permeation, *Advan. Chem. Ser.,* 1971, 118; *Chem. Eng. Med. Symp.,* 1, 1973.
10. **Cussler, E.L.,** Membranes which pump, *AICHE J.,* 17, 1300, 1971.
11. **Pedersen, C.J.,** Cyclic polyethers and their complexes with metal salts, *J. Am. Chem. Soc.,* 89, 2495, 1967.
12. **Dietrich, B., Lehn, J.-M., and Sauvage, J.-P.,** Diaza-polyoxamacrocyclic and macrobicyclic compounds, *Tetrahedron Lett.,* 2885, 1969. **Dietrich, B., Lehn, J.-M., and Sauvage, J.-P.,** Cryptates, *Tetrahedron Lett.,* 2889, 1969.
13. **Kyba, E.P., Siegel, M.G., Sousa, L.R., Sogah, G.D.Y., and Cram, D.J.,** Chiral, hinged, and functionalized multihetero-macrocycles, *J. Am. Chem. Soc.,* 95, 2691, 1973.
14. **Houk, J. and Guy, R.H.,** Membrane models for skin penetration studies, *Chem. Rev.,* 88, 455, 1988.

Chapter 2

# CHARACTERISTICS OF BIOMEMBRANE AND MEMBRANE TRANSPORT

**Naoki Kamo**

## TABLE OF CONTENTS

# I. INTRODUCTION

It is well known that cells are separated from their surroundings by a thin membrane called plasma membrane or simply biomembrane. Eucaryotic cells contain internal vesicular organelles such as mitochondria, chloroplasts, and lysosomes; these are also enveloped by membranes. The major constituents of biomembranes are lipids, proteins, and oligosaccharides which are held together by noncovalent interaction. The weight ratio of protein to lipid in most biomembrane ranges from 1:4 to 4:1 and is estimated from the specific gravity of the membrane. The thickness of biomembrane is 5 to 7 nm, and the basic structure is a bimolecular leaflet of phospholipids, the thickness of the membrane being that of the leaflet. The membrane plays several important roles in the life of the cells:

**Barrier forming a compartment** — This function of the membrane allows the cells to contain intracellular constituents and be independent of environmental changes. This provision of a compartment is essential for the autonomy and control of differential functions.

**Selective transport** — Nutrients are taken up into cells through the membrane and transported against the concentration gradient. This transport against the gradient of the concentration or activity is generally called uphill or active transport. In a biological system, such energy-requiring transport is generally called active transport, and energy comes from the metabolism. A more detailed definition of active transport will be given later. The protein machinery of an active transport system driven directly by ATP or light is sometimes referred to as a pump, especially in the field of physiology. Intracellular conditions of salt composition and concentration, pH, etc. are kept constant by virtue of the transport system in the membrane.

**Production of energy** — Bio-energy is stored and carried as a form of adenosine-3-phosphate (ATP) which is synthesized mainly by two systems: substrate level phosphorylation and oxidative phosphorylation. The latter is the major route and occurs at the membrane of mitochondria or the plasma membrane of bacteria. The ATP synthesis is essential for membrane containing an ordered array of enzymes and other proteins. When the integrity of the membrane is interrupted, the ATP synthesis is stopped even if the enzymes remain intact.

**Transformation of information** — The membrane faces the outside of the cell and is the first to receive information from outside. Receptors in the membrane convert information for the cells by biochemical reactions or a change in the membrane potential. The ability to cause the membrane potential change stems from the concentration gradient generated across the membrane by the transport. Stimuli alter the permeability of the membrane thus bringing about changes in the membrane potential. The factor which changes the permeability of the channel protein spanning the membrane at the reception of the signal is called the gate. Opening the gate changes the membrane potential and sometimes the intracellular ionic concentration which affects the metabolic reaction.

This chapter consists of four parts: (1) a brief survey of the structure and fluid nature of biomembranes; (2) transport phenomena across the membrane and definition of active transport; (3) mechanism of uphill transport through biomembranes; and (4) some examples of artificial membrane showing uphill transport.

# II. STRUCTURE AND FLUID NATURE OF BIOMEMBRANES

## A. BASIC STRUCTURE OF BIOMEMBRANE: FLUID MOSAIC MODEL

The first clue to the elucidation of the structure of biomembrane was given in 1896 by Overton,[1] who found that the rate of transport of various substances through the membrane of *Nitella,* a giant alga, correlated well with the partition coefficient between water and olive oil. He therefore suggested that lipids are important in the molecular organization of biomembrane. Lipids are relatively small molecules that have both a hydrophilic and a hydrophobic moiety. The work of Langmuir and others[2] showed that lipids can form a self-organized structure called a monolayer at the water/air interface. Lipids arrange themselves perpendicular to the surface with their hydrophilic groups directed toward the water and their hydrophobic groups toward the air. The concept of the monolayer led to the idea that lipids arrange regularly to form biomembrane. Gorter and Grendel[3] in 1925 extracted lipids from red blood cells and made a monolayer. They found that the extracted lipid occupied twice the calculated surface area of red blood cells and hence suggested that biomembranes consist of a bimolecular lipid leaflet.

Electron microscope reveals the railroad-track pictures of membranes. Since the staining reagent reacts with the polar head groups of phospholipids, the picture shows that the biomembrane has a phospholipid bilayer membrane as its basic structure. The picture also gives the membrane thickness of 5 to 10 nm. Electric capacity measurements were 0.3 to 0.5 micro $F/cm^2$, which is consistent with the bilayer structure as described below.

Phospholipids are readily self-organized to form bilayer membranes in water. When phospholipids are dispersed in saline water followed by sonication, closed vesicles are formed and the membrane is composed of a single bilayer membrane (liposome[4]). Another well-known lipid membrane is a planar bilayer membrane.[5] A membrane-forming solution such as phosphatidyl choline in decane is injected into a small hole in a partition between two aqueous media, and a lipid bilayer film is formed spontaneously. Another recently developed method now widely employed[6] is that in which a lipid monolayer is formed at two solutions separated by a plastic film with a small hole. The water levels are gradually increased and pass through the hole, and a bilayer membrane is formed spontaneously. The electric capacitance is 0.3 to 0.5 micro $F/cm^2$, which is almost equal to the calculated value assuming that thickness is 5 nm and the dielectric constant is the same as that of a hydrocarbon chain.

If the cellular surface is a lipid barrier in an aqueous environment, then the surface tension at the interface should be comparable to that of pure lipid in water, which is in the range of 10 dyne/cm. The values evaluated, however, were much less than 0.2 dyne/cm.[7] Danielli and Davson[8] proposed a model of biomembrane which depicted bimolecular leaflets of lipids and a single layer of protein covering each exposed lipid polar surface. Smaller values of the surface tension were due to the presence of surface-active proteins.

Since that model appeared, several other membrane models have been proposed. According to Singer,[9] the principles of protein structure are defined by the role of the solvent, hydrophobic interactions, hydrogen bonds, hydrophilic interactions, and electrostatic interactions. These principles are also applicable to biomembranes which are nonchemical bonding complexes between proteins and lipids. Previous membrane models did not adequately recognize these principles, but Singer and Nicolson[10] proposed a membrane model called the "fluid mosaic model" which does satisfy them.

Figure 1 illustrates this fluid mosaic model; the basic structure is bimolecular leaflets of lipids. Proteins are incorporated into the lipid bilayer in a mosaic pattern; some of them penetrate the membrane, and some strongly stuck into it. These proteins are termed "integral proteins", and their detachment requires solubilization with detergents. Freeze-fracture studies using electron microscopy[11] provided the first direct evidence for the presence of integral proteins in many biomembranes. Membranes are rapidly frozen with liquid nitrogen. The frozen membrane is fractured by the impact of a microtome knife, and the cleavage occurs along a plane in the middle of the bilayer. This technique thus provides a view of the membrane interior.

Other proteins, however, associate loosely with membranes and can be detached with mild treatment such as changing the ionic strength of the medium or by incubation with chelating agents such as EDTA; these are termed as "peripheral proteins".

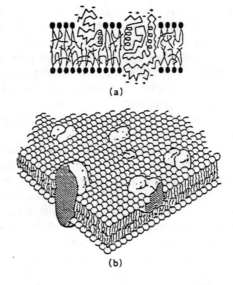

FIGURE 1.    Fluid mosaic model of biomembranes. Many proteins have the α-helix conformation. Closed circles represent the polar head group of phospholipids. (From S. J. Singer and G. L. Nicolson, *Science*, 175, 722, 1972. With permission.)

## B. ASYMMETRY OF BIOMEMBRANE

Biomembrane is asymmetric with respect to its constituents. The outer half of the membrane of human erythrocytes has phosphatidyl choline and sphingomyelin as major components, and the inner half has phosphatidylethanolamine and phosphatidylserine. Cholesterol is thought to be located mainly at the outer surface, and the sugar of oligosaccharides is also confined to this surface. The location of proteins, too, is asymmetric, and the fluid mosaic model shown in Figure 1 demonstrates their asymmetry. For example, the subunit of mitochondrial ATPase ($F_1$) binds to the membrane from the inside of mitochondria, while cytochrome c binds from the outside. What is the significance of this asymmetry, and how is it, although the membrane constituents are mobile? The answer has not yet been found. We can say, however, that asymmetry is essential for the coupling of directed flows and chemical reactions (for details, see Section IV). Examples for this coupling are ion transport by ATPase, ATP synthesis at the membrane (oxidative phosphorylation), and the proton extrusion by respiration which is the oxidation reaction.

## C. MOTION OF ALKYL CHAIN OF LIPID

The motion of alkyl chain of the lipid can be estimated by electron spin resonance (ESR) technique or fluorescence polarization. Nitroxide radical shows a three-line spectrum of ESR. The interaction of $^{14}$N-atom results in three lines of equal intensity separated by the coupling constant. The coupling constant depends on the interaction of the radical and $^{14}$N atom and shows the anisotropy. The direction of the external magnetic field with respect to the N-O axis of nitroxide affects the coupling constant as well as the g-factor, which determines the strength of the external magnetic field where resonance occurs, or the position of ESR signals in a chart. The three different spectra are obtained depending on three directions (x, y, and z axes) of applied external magnetic field with respect to the N-O axis. If the motion of nitroxide is fast enough, the signal is the average of the three spectra. The actual spectrum is a mixture of these three spectra, and the mixing ratio is dependent on the mode of motion of the radical. Analysis of the spectra gives the information on the motion of the N–O bond.[12] Nuclear magnetic resonance (NMR) is also used for the study of molecular motion.

When a fluorescent molecule is excited by polarized light, the emission light is also polarized. During the fluorescent lifetime (usually several ns), the dye rotates and the polarized plane of the emission is changed with respect to that of the excited light. Steady polarized light has been used for the excitation.[13] Recently, pulse laser light has also been implemented by which the angle of rotation of the rod-like fluorescent dye can be estimated.[14]

Results of these experiments have revealed that the motion of the carbon chain of the lipid is rapid in membranes. At low temperature, the molecular motion of the lipid molecule is restricted, as in the crystalline state. At the temperature higher than that called transition temperature, the molecular motion is increased as if in liquid. When a single kind of lipid is used, the transition is relatively narrow, suggesting that this is a cooperative process, as is phase transition. The state of high molecular motion is referred to as liquid crystal because the regular structure of the lamella remains although the molecular motion is rapid. The transition temperature of a single lipid depends on the length of the hydrocarbon chain, the number of double bonds, and the kinds of polar head groups.[15] Organisms have liquid crystal membrane at the temperature where they live: they synthesize the lipids whose transition temperature is below the prevailing temperature to maintain a highly mobile state. This is demonstrated by microorganisms.[16]

## D. FLUIDITY OF BIOMEMBRANE: LATERAL DIFFUSION

In addition to the molecular motion of the carbon chain, another important aspect of biomembrane is its fluidity: the dynamic properties and translational movements of membrane components such as lipids and proteins. These properties are demonstrated by various bio-physical and biochemical methods. McConnell[17] and Träuble[18] showed the lateral diffusion of phospholipids in a bilayer membrane using ESR. The radical of nitroxide shows the three line spectrum of ESR described above, but at high concentration, a spin-spin exchange occurs, and the signal becomes broad. If the spin-spin exchange is extremely rapid, the signal shows only one peak instead of three. Thus, analysis of the shape of the ESR spectrum provides information on the interaction between radicals.

Nitroxide-spin labeled phospholipids were applied as a spot to the center of the lipid layer formed on a quartz plate. At the start of the experiment, only one peak of the ESR signal was observed, indicating a strong spin-spin exchange, but with the passage of time, the three-line spectrum appeared. At an earlier stage, the width of the spectrum was large; this became sharper with time, indicating that the labeled phospholipids diffused and the spin-spin interaction weakened as time passed. Detailed analysis of the change of the signal shape and intensity gave the diffusion constant of the lipid in the membrane.

Lateral diffusion was visualized by Frye and Edidin,[19] who observed the movement of the fluorescent dye attached to protein under a fluorescent microscopy. Green and red fluorescing dyes were respectively attached to membrane proteins of mouse and human cells using an antigen-antibody reaction. The two types of cells were fused by hemagglutinating virus of Japan. Immediately following fusion, the red-fluorescing protein occupied half of the fused cells, and the green-fluorescing protein the other half. After 40 min at 37°C, the red and green fluorescence had mixed together. This process required no energy (ATP hydrolysis), and no effect of a reagent inhibiting protein synthesis was observed, indicating that the mixture was formed by the diffusion of proteins.

More quantitative methods for measuring lateral diffusion are fluorescence photobleaching recovery (FPR) and fluorescence correlation spectroscopy (FCS). A target protein is first labeled with a fluorescent chromophore. When FPR is employed, strong light from a laser bleaches the pigment. Immediately thereafter, the laser light is weakened in order to measure the amount of pigment at the bleached spot. Observance of the recovery of the fluorescence means that dye-labeled proteins have come into the spot from the surrounding areas. The time course of the fluorescence recovery gives the diffusion constant of the protein. The

FCS method measures the fluctuation of the fluorescence intensity at a tiny spot. The fluctuation stems from the lateral random-walk of the labeled protein: sometimes this protein enters and exits from the spot being observed.

These measurements indicate that the diffusion constant of proteins in biomembrane, D, is about $10^{-9}$ to $10^{-11}$ cm²/S. Note that D of small molecules in water is the order of $10^{-5}$. Calculation by the Einstein-Stokes law yields a 'viscosity' estimated to be several of poise, which is almost equivalent to that of glycerol or olive oil. Thus, in a first approximation, a membrane is viewed as a two-dimensional solution of protein in lipid. Recently, however, proteins have not always been freely mobile, and the mobility of some is regulated by interactions between membrane proteins and proteins located on the membrane surface or in the cytoplasm adjacent to the inner membrane surface (the cytoskeleton). We must also note that a labeled molecule is not the same as a native one. For example, the spin-labeled reporter molecule is usually bulky, and then the interaction between the surroundings may be changed.

## E. "FLIP-FLOP" OF PHOSPHOLIPID

Kornberg and McConnell[21] showed that lipid molecules can move from one side of the liposome bilayer to the other side, and this movement is referred to as "flip-flop", to distinguish it from in-plane translational mobility (lateral diffusion). The time interval for 50% of the molecules to flip from one monolayer to the other ranges from hours to days or even weeks. The slowness is caused by the polar head groups of lipids having to go through the hydrophobic portion in the middle of the membrane, a movement which takes a great deal of energy. Thus the biological significance of this flip-flop is not clear. Since proteins are much larger than lipids, trans-membrane rotation of the membrane is improbable. Therefore, the idea that the substance will be transported by cross-membrane rotation (see carrousel in Figure 8) is also unlikely.

## III. TRANSPORT PHENOMENA IN BIOMEMBRANES

## A. SIMPLE AND FACILITATED DIFFUSION

Transport through membranes is classified as follows:

1.   Simple diffusion
2.   Facilitated diffusion
3.   Active transport
     a.   Primary active transport
     b.   Secondary active transport
4.   Group translocation
5.   Cytosis

Simple diffusion is the downhill movement of solute in accordance with the gradient of electrochemical potential of the solute. Flux, J, is defined as mol/cm²/S and expressed as follows:

$$J = -uCd\mu/dx$$

Here, $d\mu/dx$ is the gradient of electrochemical potential and

$$\mu = \mu^\circ + RT\ln a + zF\psi$$

where $\mu^\circ$, a, and $\psi$ are the standard chemical potential, activity, and electrical potential. R,

T, z, and F stand for the gas constant, Kelvin temperature, valence of solute, and Faraday constant, respectively. Activity, a, is usually assumed to be equal to concentration for the sake of convenience. If this assumption is taken and if the solute is electrically neutral, the flux equation can be recast to

$$J = -(uRT/F) \, dC/dx = -D \, dC/dx$$

where D is defined as uRT/F which is called diffusion coefficient.

Since the barrier is usually the membrane, the transport within the membrane is rate-determining, so that the rate of diffusion is determined by the concentration gradient within the membrane. On the other hand, the observable motive force for the diffusion is the concentration difference in two aqueous phases separated by the membrane. Therefore, we obtain

$$J = -P\Delta C$$

The proportionality coefficient, P, is known as the permeation coefficient and includes the partition coefficient between the concentrations of aqueous phase and that within the membrane. The discovery by Collander et al. and Overton[1] that the rate of permeation of varying compounds through the membrane of *Nitella,* a giant plant cell, which is well correlated with the partition coefficients between olive oil and aqueous phase, has this relationship as its basis.

The facilitated transport is that accomplished with the aid of a carrier within the membrane, and the transport direction is forward to the chemical potential as is true of the simple diffusion. The presence of the carrier leads to specificity of the solute to be transported. For example, glycerol shows the permeability coefficient of $10^{-7}$ cm/S for the membrane of plant cells or bovine red cells. On the other hand, the coefficient of human red cells is $10^{-5}$ cm/S, which is larger than those of plant cells or of bovine red cells. Ethyleneglycol inhibits the glycerol transport. This phenomenon is easily interpreted by assuming that there exists a carrier (denoted as X) and that a complex (SX) between the carrier and solute (S) is transported within the membrane:

$$S + X \rightleftharpoons SX \rightarrow transport$$

Therefore, the rate of transport is expressed as it is for the enzymatic reaction:

$$V = V_{max} \, [S]/(Km + [S])$$

where $V_{max}$, [S], and Km are the maximum rate of the transport, solute concentration, and the solute concentration where $V = (1/2)V_m$, respectively. This equation means that the rate of transport may be saturated at a high concentration of [S], while the rate of simple diffusion is proportional to [S].

## B. ACTIVE TRANSPORT

One important function of the biomembrane is active transport, which is transport against the concentration gradient. By virtue of this function, many organisms can live in a nutrient-poor environment such as a river or a lake. Active transport also plays a very important role in the human body: the concentration of $Na^+$ inside cells is lower than that of outside cells while the concentration gradient of $K^+$ is the opposite. $Na^+$ or $K^+$ may flow into or out of a cell according to the concentration gradient: hence, the maintenance of the concentration gradients across the cell membrane reflects the existence of the transport system that pumps

Na$^+$ out and takes K$^+$ into cells. This process is required for the energy. About 70% of basic metabolic energy is used for this Na/K transport. The Na$^+$ gradient is used for the active transport of various nutrients such as sugars and amino acids; nerve excitation also requires this concentration gradient. For muscle contraction and information transmission, Ca$^+$ transport is essential. Ca$^{2+}$ is actively accumulated to storage vesicles, where signals then release the ion to evoke the function. Thus, transport across membranes is crucial to the maintenance of the life of organisms.

## C. DEFINITION OF ACTIVE TRANSPORT

Rosenberg[22] gave a definition of the active transport as transport against the electrochemical potential of the solute. Using[23] verified that the sodium transport through the frog abdominal skin is compatible with Rosenberg's definition. When the skin is mounted between two chambers containing oxygenated and stirred Ringer's solution, the skin generates an electrical potential difference across itself; the inside is electrically positive with respect to the outside. A pair of electrodes is immersed in two chambers across the skin and an electric current (denoted as I$_{sc}$) is passed which cancels the potential difference. In this situation, since the composition of the two solutions separated by the skin is identical, there is no electrochemical potential gradient of the components between the two solutions. Using radioactive isotopes, the fluxes of Na$^+$ and Cl$^-$ can be determined. When $^{22}$Na and $^{24}$Na were used simultaneously, the Na flux from outside to inside, J$_{Na}^{in}$, and Na flux of the reversed direction, J$_{Na}^{out}$, were determined in the same preparation. Even under a condition where there is no electrochemical potential gradient, J$_{Na}^{in}$ is much larger than J$_{Na}^{out}$, while J$_{Cl}^{in}$ is equal to J$_{Cl}^{out}$. This means that Na transport is active transport in accordance with Rosenberg's definition. Furthermore, if the skin is poisoned (e.g., with CN$^-$) or the oxygenation of the bathing solution is stopped, the inward and outward flux of Na$^+$ become equal, showing the requirement of energy.

From the standpoint of irreversible thermodynamics, the entropy production of the system must increase. For details of entropy production, see the books[24] cited. It is not important that each flow in the system necessarily increase its entropy production, only that the sum of entropy production of all flows increases. Let us assume that there are two simultaneous flows (a flow contains a chemical reaction) in the system referred to as A and B. The entropy production of flows A and B is denoted as $\Phi_A$ and $\Phi_B$, respectively. Thermodynamics states that $\Phi_A + \Phi_B$ should be positive, but the case where one of the production is negative is allowed as long as the total sum is positive. Negative entropy production means that the process is thermodynamically uphill and is not allowed to occur naturally if a system contains this flow alone. The thermodynamically downhill process of B supports the thermodynamically uphill process of A.

Kedem[25] proposed a definition of active transport on the basis of irreversible thermodynamics. According to irreversible thermodynamics, the relation between fluxes and forces is in terms of the following phenomenological equations, where the flows are written as linear functions of the forces:

$$J_1 = L_{11}X_1 + L_{12}X_2 + \ldots\ldots + L_{1r}X_r$$

$$J_i = L_{i1}X_1 + L_{i2}X_2 + \ldots\ldots + L_{ir}X_r$$

where $X_i$ is the i-th force and $L_{ij}$ is the phenomenological coefficient. The suffix r represents the chemical reaction. Rearrangement of these phenomenological equations leads to

$$J_i = R_{ii}X_i + \Sigma R_{ij}J_j + R_{ir}J_r$$

The first term on the right-hand side is the flow of i-th component by its own force, and the second term represents the contribution of the flux of j-th component. In other words, it is the flow of i-th component caused by the flow of other components. The last term is the flux of i-th component caused by the chemical (metabolic) reaction. Kedem defined the transport whose $R_{ir}$ is not zero as the active transport.

Active transport is usually classified into two groups: primary and secondary active transport. Primary active transport is that based on the third term on the right-hand side of the above equation. In other words, it is the active transport which is in exact accord with Kedem's definition. Metabolic energy (in many cases, hydrolysis of ATP) converts directly to the transport energy, and this conversion is done at a single or complex of proteins.

Secondary active transport is the transport based on the second term of the above equation. The downhill transport of another ion (denoted as A) whose gradient is generated by metabolic energy (i.e., by the primary active transport) induces the uphill transport of the substance in question (denoted as S). Two cases are possible, depending on the direction of the transport of A and S. When the directions of transport of A and S are the same, it is called symport or co-transport. The transport protein mediating the symport is referred to as a symporter. When the directions of the transport of A and S are opposite, the transport is called antiport, and the membrane protein mediating the antiport is antiporter. A uniport system is also of record: the ion in question is electrophoretically transported by the membrane potential generated by the metabolic reaction. The machinery that mediates the symport, antiport, and uniport is termed a transporter. Figure 2 illustrates these transport phenomena.

## D. EXAMPLES OF THE PRIMARY ACTIVE TRANSPORT

Cells generally show a high intracellular level of potassium ion and a low intracellular level of sodium ion, whereas the external environment of these cells shows a converse pattern. In animal cells, this gradient can be kept by virtue of a membrane-bound enzyme called $Na^+,K^+$-ATPase.[26] This enzyme hydrolyzes ATP only in the presence of both $Na^+$ and $K^+$. Physiologically, the membrane device which mediates the uphill $Na^+$-transport is called an "Na-pump". The conclusion that $Na^+,K^+$-ATPase is the Na-pump has been drawn because of the parallelism between the enzymatic activity and transport activity under various conditions.

This transport system has been examined in detail for the red blood cell[27] and squid giant axon.[28] "Ghosts" are red cells that have lost most of their hemoglobin as well as other intracellular constituents by osmotic hemolysis. Such emptied red cells retain a high degree of membrane integrity and continue to show osmotic phenomena. When resealing occurs in the presence of nonpermeating substances such as ATP and glucose, these substances are entrapped inside the cell. Cells entrapping this type of energy source can then maintain the Na/K concentration gradient, whereas cells without an energy source cannot. Because of the large size of the giant axon of the squid, a cannula can be used to infuse an energy source such as ATP or phosphoarginine directly into the cell. $Na^+$ flux is measured using $^{22}Na$. Poisoning of the cell with cyanide largely inhibited the sodium-ion extrusion, but this is restored with subsequent infusion of ATP or phosphoarginine. The $Na^+$ and $K^+$ transport is therefore energy-dependent.

With the hydrolysis of 1 mol of ATP, 3 mol of $Na^+$ and 2 mol of $K^+$ are transported. This value was first obtained using red blood cells. By means of this stoichiometry, we can calculate the efficiency of the energy conversion. The concentrations of $Na^+$ and $K^+$ inside or outside red blood cells and the membrane potential are $[Na^+]_{in} = 10$ m$M$, $[Na^+]_{out} = 145$ m$M$, $[K^+]_{in} = 150$ m$M$, and $[K^+]_{out} = 5$ m$M$; the membrane potential, $\Delta\psi$ (defined as the interior electrical potential with respect to the outside), is $-8.6$ mV, which is approximately equal to the equilibrium potential of $Cl^-$. The energy required for the extrusion of 3 mol $Na^+$ is

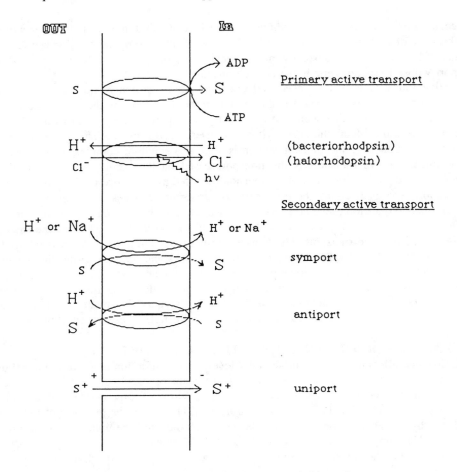

FIGURE 2.    Classification of active transport in biomembrane.

$$\Delta G_{Na} = 3[\mu_{Na}(out) - \mu_{Na}(in)]$$

$$= 3RTln(145/10) - 3F(-0.0086)$$

The energy of the uptake of 2 mol $K^+$ is

$$\Delta G_K = 2[\mu_K(in) - \mu_K(out)]$$

$$= 2RTln(150/5) + 2F(-0.0086)$$

Therefore, the free energy for the transport, $\Delta G_t$ is

$$\Delta G_t = \Delta G_{Na} + \Delta G_k = 9.31 \text{ kcal/mol}$$

When we assume that [ATP] = 1.5 m$M$, [ADP] = 0.32 m$M$, and [Pi] = 0.32 m$M$, the free energy change by the hydrolysis of 1 mol ATP is calculated to be

$$\Delta G_p = \Delta G_{po} + RTln[ATP]/[ADP][Pi] = -13.0 \text{ kcal/mol}$$

where $\Delta G_{po}$ is taken as 7.3 kcal/mol. The values of $\Delta G_t$ and $\Delta G_p$ show the high efficiency of energy conversion.

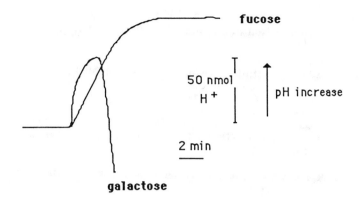

FIGURE 3. β-galactoside is taken up by *E. coli* cells with $H^+$-symport. Starved cells were suspended in a weakly-buffered medium to measure the pH change. At the time shown by the arrow, galactose or fucose was added and upward deflection corresponded to the alkalization in the medium (i.e., uptake by cells).

Other examples of the primary active transport are $Ca^{2+}$-ATPase, $H^+$-ATPase, and $H^+,K^+$-ATPase. $Ca^{2+}$-ATPase works as a calcium pump in the membrane of the sarcoplasmic reticulum to accumulate ions. The sarcoplasmic reticulum membranes are the main constituents of muscle's intracellular membranes. The rapid activation and inactivation cycle of muscle depends on the sudden release of calcium ions and their subsequent complete removal. The removal is done by accumulation to the sarcoplasmic reticulum fueled by ATP. The electrical signal releases the calcium from the reticulum. The pH in the stomach is very low, and proton is transported by $H^+,K^+$-ATPase in the membrane.

$Na^+,K^+$-ATPase and $Ca^+$-ATPase are reversible. When the ion-gradient is imposed so that the direction of ion-diffusion might be reversed, for example, the $Na^+$-transport may occur from outside to inside of red blood cells and that of $K^+$ might be opposite: the enzyme can synthesize ATP from ADP and phosphate. $H^+$-ATPase is known to be the enzyme in the membrane of mitochondria or bacteria which catalyzes the ATP synthesis by an expenditure of $H^+$-gradient across the membrane or generates $H^+$ gradient by an expenditure of ATP. Recently, $H^+$-ATPase has been found in lysosome[29] or in vacuole[30] membrane of plant cells where it maintains the acidity of the interior of these membrane systems.

One of the bacteria living in high NaCl medium (ca. 4.25 *M*) is *Halobacterium halobium*. This bacteria has a membrane protein called bacteriorhodopsin which works as a light-driven $H^+$-pump.[31] Illumination generates the transmembrane gradient of proton chemical potential. This may be classified as primary active transport, although the energy source is not chemical energy but light.

## E. EXAMPLES OF THE SECONDARY ACTIVE TRANSPORT

Figure 3 demonstrates that *Escherichia coli* cells accumulate β-galactoside by the antiport of $H^+$. Starved cells were suspended in weakly buffered solution, the pH of which was monitored with a glass electrode. When the pH stabilized, galactose was added. Alkalization occurred, followed by acidification. The acidification phase then disappeared, and only the alkalization phase could be observed when fucose, a nonmetabolic sugar, was used or when a mutant lacking for enzymes of galactose metabolism was used. Therefore, the initial alkalization is due to the entry of $H^+$ via the symporter, and the subsequent acidification is the result of the metabolic production of acidic substances.

A number of bacterial transport systems have symport or antiport systems with $H^+$ movement.[32] The driving force for the $H^+$-movement is the difference of proton electrochemical potential across the membrane.

## TABLE 1
### Comparison of Δp and 59log[TMG]$_{in}$/[TMG]$_{out}$

| Source of energy | ΔE (mV) | ΔpH | Δp (mV) | [TMG]$_{in}$/[TMG]$_{out}$ | 59log[TMG]$_{in}$/[TMG]$_{out}$ (mV) |
|---|---|---|---|---|---|
| Glucose | −35.5 | 0.75 | −79.9 | 16.1 | 71.2 |
| Arginine | −39.0 | 0.0 | −39.0 | 5.0 | 41.2 |

*Note:* The finding the sum of Δp and 59log[TMG]$_{in}$/[TMG]$_{out}$ is approximately equal to zero implies that the ratio of H$^+$ and TMG is 1:1.

Data from Kashket, E. R. and Wilson, T. H., *Biophys. Res. Commun.*, 59, 884, 1974.

Proton electrochemical potential is written as

$$\mu_{H^+} = \mu_{H^+}^{\circ} + RTlna_{H^+} + F\psi$$

$$= \mu_{H^+}^{\circ} - 2.3RTpH + F\psi$$

The difference with respect to the outside is

$$\mu_{H^+}(in) - \mu_{H^+}(out) = -2.3RT\Delta pH + F\Delta\psi$$

where $\Delta\vartheta$ is defined as $\vartheta(in) - \vartheta(out)$. Division with F of the above equation yields:

$$\Delta p = [\mu_{H^+}(in) - \mu_{H^+}(out)]/F$$

$$= -59\Delta pH + \Delta\psi(mV)$$

This quantity is defined as Δp and called proton motive force,[35] whose unit is that of electric potential, usually mV.

Table 1 shows the steady state accumulation ratio of thiomethyl-β-galactose (TMG), nonmetabolizing sugar, and Δp. When the stoichiometry of symporter is that n mole of sugar (denoted as S) taken up by expenditure of H$^+$ uptake of n mole, the following equation may be held (assuming 100% of efficiency).

$$n[\mu_s(in) - \mu_s(out)] + m[\mu_{H^+}(in) - \mu_{H^+}(out)] = 0$$

As shown in Table 1, m:n = 1:1. Another well recognized example of symport is the uptake of sugars in the intestines which is driven by Na$^+$ gradient.

Other examples of an antiport system are Na$^+$/H$^+$, K$^+$/H$^+$, and Na$^+$/Ca$^{2+}$ antiport. In bacterial cells, Na$^+$/H$^+$ antiporter extrudes Na$^+$ and the driving force is Δp which is generated by the electron transport on the membrane. In other words, oxidation of substances at the electron transport chain makes the gradient of the proton electrochemical potential, Δp, which creates Na$^+$-gradient. In animal cells, on the other hand, Na$^+$ extrusion is due to Na$^+$,-K$^+$-ATPase, which is driven directly by the chemical energy.

If $\mu_{H^+}$ or $\mu_{Na^+}$ is the driving force in bacterial cells, an artificially imposed driving force must stimulate the transport. The membrane potential can be artificially generated by addition of valinomycin to the membrane which separates the potassium gradient. Indeed, the transport can be observed when the potential is imposed artificially.[34]

## F. MECHANISM OF ACTIVE TRANSPORT

The active transport is composed of the following four steps: (1) recognition of solute (binding of solute to carrier), (2) translocation through the membrane, (3) energy transduc-

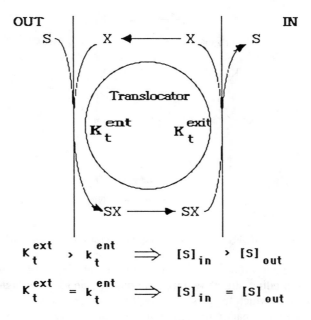

FIGURE 4.   Kinetic analysis of active transport system.

tion, and (4) dissociation of solute from carrier. Isolation of some of the carrier proteins (translocator) and determination of their primary structure are now in progress, and in the future, the molecular events involved in solute transport will be clarified.

Kinetic analysis has revealed that the energy changes the dissociation or affinity constants of the solute to the carrier inside and outside of the membrane. The dissociation constant at the outside interface is represented by $K_t^{ent}$ and that at the inner surface by $K_t^{ext}$. The rate of transport from outside to inside is formulated as follows, provided the binding of solute to the carrier is rapid and the rate determining step is the transport across the membrane.

$$V = V_{max}[[S]_{out}/([S]_{out} + K_t^{ent}) - [S]_{in}/([S]_{in} + K_t^{ext})]$$

When the steady state is reached, $V = 0$, and this relation gives the following equation.

$$[S]_{in}/[S]_{out} = K_t^{ext}/K_t^{ent}$$

This equation means that

$$\text{if} \quad K_t^{ext} > K_t^{ent} \quad \text{then} \quad [S]_{in} > [S]_{out}$$

and

$$\text{if} \quad K_t^{ext} = K_t^{ent} \quad \text{then} \quad [S]_{in} = [S]_{out}$$

In other words, this means that the change in the dissociation constant leads to the accumulation (see Figure 4). This relation is shown to hold in many transport systems. Winkler and Wilson[35] measured $K_t^{ext}$ and $K_t^{ent}$ of lactose/$H^+$ symport in *Escherichia coli* when cells were energized or when cells were poisoned. $K_t^{ent}$ was not energy-dependent while $K_t^{ext}$ was changed by a factor of $10^2$ by supply of energy.

Figure 5 shows a highly simplified model of $Na^+,K^+$-ATPase.[36] The enzyme has two

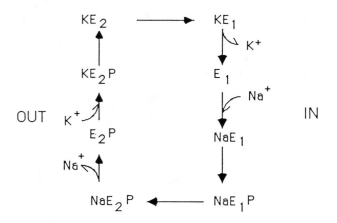

FIGURE 5.    Highly simplified model of ion transport by $Na^+,K^+$-ATPase.

forms of $E_1$ and $E_2$: the ion-binding site of $E_1$ is located inside, and $E_1$ has high affinity for $Na^+$; while the ion-binding site of $E_2$ faces the outside, and $E_2$ has high affinity for $K^+$. $Na^+$ binds the enzyme, which triggers the phosphorylation by ATP ($NaE_1P$), and phosphorylation changes the conformation to $E_2$. $E_2P$ has low affinity for $Na^+$, which is released to the outside. In contrast, $E_2P$ has high affinity for $K^+$ so that $K^+$ on the outside binds to the enzyme ($KE_2P$). The binding triggers the dephosphorylation to $KE_2$, which is not stable and converts to $KE_1$ form. This $E_1$ form has low affinity for $K^+$ but has high affinity for $Na^+$, and therefore, an $Na^+/K^+$ exchange occurs.

The $Ca^{2+}$-ATPase cycle resembles the $Na^+,K^+$-ATPase cycle in several points.[37] The enzyme also has $E_1$ and $E_2$ forms. $E_1$ form opens a binding cavity to the outside (cytosol *in vivo*) and $Ca^2$-binding site of $E_2$ form faces the inside. Phosphorylation favors $E_2$, and dephosphorylation favors $E_1$. Transition of the conformation greatly changes the binding affinity for $Ca^{2+}$. According to these models, the difference in affinity for ions between the $E_1$ and $E_2$ forms is essential in the transport mechanism. The molecular mechanism of the conformation change is not yet known, but the change in the binding affinity drives the ion movement across the membrane.

Halorhodopsin[38] (hR) is a light-driven, inwardly-directed chloride pump in the membrane of *Halobacterium halobium*, which lives in lakes or pools from which salt is produced. The culture medium of bacterium contains $4.3\ M$ NaCl. This bacterium also has bacteriorhodopsin in its membrane, and this pigment functions as a light-driven proton pump. These pigments have retinal, vitamin A aldehyde as a chromophore. On illumination, *all-trans* retinal isomerizes to 13-cis retinal, and the protein conformation subsequently changes. Several photointermediates are found which are linearly aligned and cyclic: photoexcitation initiates a cycle of conformational change to the original hR, called a photocycle. The photocycle of halorhodopsin is quite interesting with respect to its mechanism of transport (Figure 6). The suffix numbers indicate the wavelength of absorption maximum. In the dark, hR exits in chloride-bound form, $hR\text{-}Cl_{578}$, since the medium contains $4\ M$ NaCl, and the Km value of the chloride-dependent equilibrium between $hR_{578}$ and $hR_{565}$ is approximately equal to 20 m$M$. Photoexcitation of $hR\text{-}Cl_{578}$ becomes an intermediate of $hR\text{-}Cl_{520}$ via an earlier intermediate whose lifetime is in the micro-second time range. There exists another chloride-dependent equilibrium between $hR\text{-}Cl_{520}$ and $hR_{640}$, but its Km value is much larger than the former equilibrium. The difference in Km value between these two equilibria creates the chloride transport provided that the $Cl^-$-binding cavity at the equilibrium between $hR_{565}$ and $hR\text{-}Cl_{578}$ faces outside and, that at another equilibrium, faces inside. Bacteriorhodopsin also has intermediates which lose or regain $H^+$.

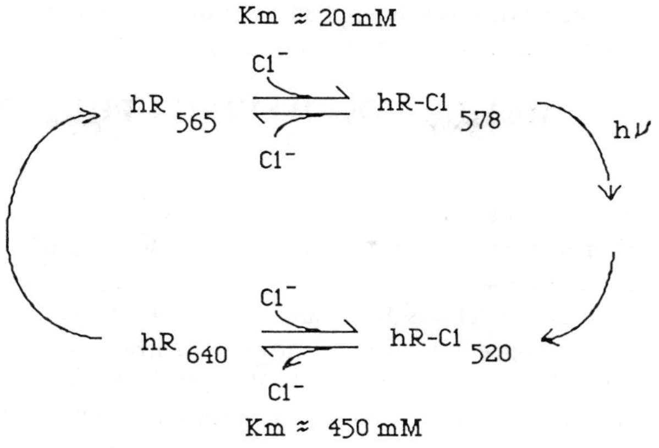

$$Km \approx 20\,mM$$

FIGURE 6. Photocycle of hR.

The uptake of sugars or amino acids in the intestines requires the presence of $Na^+$ in the medium.[39] Kinetic analysis revealed the following scheme: $Na^+$ in the medium interacts with carrier X to form a complex $XNa^+$. This binding changes the conformation of X to increase the binding affinity, and a triple complex of X, $Na^+$, and S is formed. The complex is moved to another surface, and S and $Na^+$ are released to the inside, although the mechanism of the movement within the carrier protein in the membrane is not yet understood.

$$S + Na^+ \rightleftharpoons NaX \qquad \text{binding constant} \quad K_1$$

$$NaX + S \rightleftharpoons NaXS \qquad \text{binding constant} \quad K_2$$

The transport rate is formulated as

$$V = V_{max}\{[S]/([S] + Km)\}$$

where $Km = 1/K_1K_2[Na] + 1/K_2$, indicating that the Km value is a function of $Na^+$. The asymmetry of $Na^+$ concentration changes the affinity of carrier to the solute. If we assume that S first binds to X and then $Na^+$ binds to the binary complex, SX, a different kinetic equation is obtained.

From these examples, we may conclude that the asymmetry of the binding affinity is essential irrespective of primary and secondary active transport. The mechanism of the transfer within the membrane is unknown although it may be made clear in the near future. Since, as described above, the flip-flop motion or cross-membranous rotation is not possible, the solute must pass through the carrier protein.

## G. GROUP TRANSLOCATION

A typical example of group translocation is the uptake of glucose by bacterial cells.[40] Glucose is phosphorylated during its transport through the membrane. The transport in which a transporting solute is modified is called group translocation. The ionized sugar (phosphorylated sugar) does not easily leak out, and the first step of the glucose metabolic pathway is completed during transport. In *Escherichia coli* and *Salmonella typhimurium*, glucose, fructose, or mannose is transported by group translocation, and lactose, maltose, pentose, or galactose is transported by the symport with $H^+$. In *Staphylococcus aureus* all sugars are transported by group translocation.

## H. CYTOSIS

Cytosis is a transport which is accompanied by deformation of the membrane. When liquid is taken, the term pinocytosis is applied, and the case for solid particles is called phagocytosis. These phenomena are often seen in amoebic cells.

## IV. ARTIFICIAL MEMBRANES HAVING UPHILL TRANSPORT ACTIVITY

Other chapters in this book provide a description of various forms of uphill or active transport in artificial membrane, so only a bird's eye view of it will be given here. As emphasized in the active transport of biomembranes, one essential feature is the asymmetry of binding affinity of the carrier. This is true in artificial membrane which has uphill transport. The asymmetry creates a concentration gradient of the carrier-solute complex within the membrane, and the direction of the flow due to this gradient is opposite that anticipated from the concentration gradient of the two solutions separated by a membrane.

The first publication on artificial active or uphill transport membrane appeared as early as 1969.[41] The membrane is constructed by dissolving the carrier into water-immiscible organic solvent. The carrier used is fatty acid which gains binding ability for cation in alkaline solution, but loses it in acidic solution due to association and dissociation of the carboxyl group. Cation is therefore transported from alkaline to acidic solution against its own concentration gradient. The downhill transport of proton through the membrane drives the uphill transport of cation, or we can say that trans-membranous neutralization drives the transport.

The selectivity of solute to be transported is made[42] using monensin, which is a transporting antibiotic for $Na^+$. In alkaline solution, monensin forms a ring into which $Na^+$ is incorporated; while in acidic solution, it does not have this ability. The uphill transport of $Na^+$ from an alkaline to acid compartment has been documented.

The antiport of electrons and a negative solute was also reported.[43] Ferrocene or TMPD (N,N,N',N'-tetramethyl-p-phenylenediamine) was used as carrier and dissolved in an organic solvent which separated two solutions of different redox potential. When ferrocene is oxidized it bears a positive charge, and thus it works as a carrier for negatively charged ions. Oxidation of ferrocene takes up hydrophobic anions into the organic membrane. In contrast, the reduced form is neutral and reduction of the complex of positively charged ferrocene and anion is released into the solution. Cycles of the oxidation and reduction lead to uphill transport of lipophilic anion (picrate or $ClO_4^-$) from the solution containing oxidizing agents to the solution containing reducing agents. Even when two solutions were shortened with a pair of calomel electrodes, the transport rate was not changed, meaning that this is active transport according to the criteria of Using and Rosenberg. The amounts of anion transported and those of trans-membrane reaction were compared and agreed well with each other. Moreover, the transport drives the cross-membranous redox reaction. These facts indicate that the coupling between the transport and the reaction is tight. In combination with a suitable carrier for cations, trans-membrane redox reaction can move cations.

An organic liquid membrane containing lipophilic quaternary ammonium which serves as a carrier separates the two alkaline and acidic solutions.[44] Uphill transport of amino acids from alkaline to acid solution was observed, and the mechanism of the transport is different from those described above. The binding affinity of the carrier is not changed, but the ionic states of the solute are. In alkaline solution, amino acid exits as anion and is then extracted to the organic membrane with quaternary ammonium cation. But, in acidic solution, amino acid is changed to anion, which gives rise to the release of amino acids into the solution. This is a kind of group translocation in the sense that the chemical form of the solute is changed.

The following is another example[45] of group translocation using the enzymes, hexokinase, and phosphatase which catalyze the reactions:

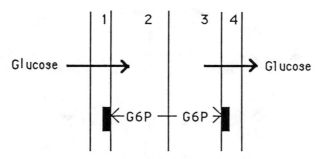

1: Cellophane (negatively charged)
2: Hexokinase membrane
3: Phosphatase membrane
4: Cellophane (negatively charged)

FIGURE 7. Example of an active transport membrane system using an enzyme. Thick lines indicate that G6P does not pass through.

Glucose + ATP → Glucose-6-phosphate (G6P) + ADP    (hexokinase)

G6P → Glucose + Phosphate    (phosphatase)

Figure 7 shows the construction of a membrane system which consists of (1) negatively charged cellophane membrane, (2) hexokinase-immobilized membrane, (3) phosphatase-immobilized membrane, and (4) negatively charged cellophane membrane. Glucose in the left compartment enters the hexokinase membrane where it is decomposed as described above. Since the G6P formed is negatively charged molecules, it cannot diffuse toward the left in the figure, but only toward the phosphatase membrane. Because of the cellophane membrane (4), G6P is forced to stay in the enzyme layer where the enzymes synthesize glucose. The glucose can pass through the membrane into the right compartment. The overall reaction is written as

Glucose(left) + ATP → Glucose(right) + ADP + Phosphate

The hydrolysis of ATP apparently drives the glucose transport. Chemical reaction, in general, cannot create a one-directional flow in an isotropic system, but such a flow can be created by restricting the movement of a chemical (G6P) in a certain direction using a barrier (cellophane membrane). Asymmetry of the membrane system is necessary to create the flow which is caused from chemical reactions, in accord with the Curie principle.[24]

## V. CONCLUDING REMARKS

The possible transport types are shown in Figure 8. Transport within artificial membranes which have uphill or active transport activity is resulted from a downhill diffusion within membranes, as described above. On the contrary, the carrier protein in biological membranes is usually integral protein which is deeply embedded in or traversing the lipid bilayer of membranes. Although the biomembrane is fluid, no flip-flop or transverse rotation of the protein occurs. The solute is transported by a channel mechanism, but the key factor in translocation of the solute against the concentration may be the rotation of a segment of the carrier protein. How does the fluid nature of biomembrane contribute to the enhancement of the transport activity? The answer may be that amino acid residues in the protein are in a state of high molecular motion.

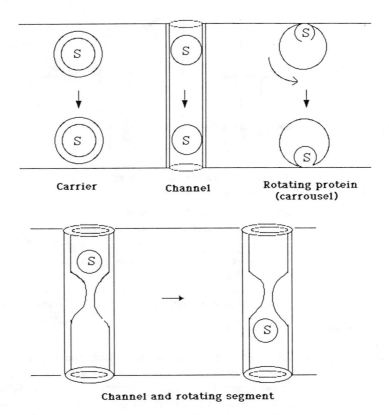

FIGURE 8.    Possible types of transport mechanism.

Almost all of the artificial transport system is carrier-type where the complex between the carrier and the solute is diffused in accordance with the chemical potential within the membrane, and this transport is uphill with respect to the concentration of the two solutions separated by the membrane. Synthesis of a channel-type carrier with uphill transport may be a very interesting challenge to investigate. The success in constructing artificial channel type carriers or translocators and analysis of their way of functioning may, in turn, contribute greatly to an understanding of the molecular mechanism of carrier proteins.

Other aspects to be considered for understanding the transport through biomembranes may be the specificity of transporting solute. In biological systems, the stereospecificity is rigid and originates from the simultaneous recognition of three or more features of substrate structure in space. Host molecules having more than one recognition site may act as a carrier molecule which is highly selective for a certain substance.

# REFERENCES

1. **Stein, W.D.,** Permeability for lipophilic molecules and **Sha'afi, R.I.,** Permeability for water and other polar molecules, in *Membrane Transport,* Bonting, S.L. and de Pont, J.J.H.H.M., Eds., Elsevier, Amsterdam, 1981.
2. **Davies, J.T. and Rideal, E.K.,** *Interfacial Phenomena,* Academic Press, New York, 1961.
3. **Gorter, E. and Grendel, F.,** Spreading of oxyhemoglobin, *J. Exp. Med.,* 41, 439, 1925.
4. **Huang, C. and Thompson, T.E.,** Preparation of homogeneous, single walled phosphatidylcholine vesicles, in *Methods in Enzymology,* Vol. 32, Fleischer, S. and Packer, L., Eds., Academic Press, New York, 1974, 485.

5. **Hanai, T., Haydon, D.A., and Taylor, J.,** Some further experiments in bimolecular lipid membranes, *J. Gen. Physiol. Suppl.,* 48, 59, 1965. **Huang, C. and Thompson, T.E.,** Properties of lipid bilayer membranes separating two aqueous phases, determination of membrane thickness, *J. Mol. Biol.,* 13, 183, 1965. **Mueller, P., Rudin, D.O., Tien, H.T., and Wescott, W.C.,** Methods for the formation of single bimolecular lipid membranes in aqueous solution, *J. Phys. Chem.,* 67, 534, 1963.

6. **Montal, M. and Mueller, P.,** Formation of bimolecular membranes from lipid monolayers and a study of their electrical properties, *Proc. Natl. Acad. Sci. U.S.A.,* 69, 3561, 1972.

7. **Cole, K.S.,** Surface forces in the arbacia egg, *J. Cell. Comp. Physiol.* 1, 1, 1932.

8. **Danielli, J.F. and Davson, H.,** A contribution to the theory of permeability of thin films, *J. Cell. Comp. Physiol.,* 5, 495, 1935.

9. **Singer, S.J.,** The molecular organization of biological membranes, in *Structure and Function of Biological Membranes,* Rothfield, L.I., Ed., Academic Press, New York, 1971.

10. **Singer, S.J. and Nicolson, G.L.,** Fluid mosaic model of the structure of cell membranes, *Science,* 175, 722, 1972.

11. **Branton, D.,** Fracture faces of frozen membranes, *Proc. Natl. Acad. Sci. U.S.A.,* 55, 1048, 1966.

12. **Hubbell, W.L. and McConnell, H.M.,** Molecular motion in spin-labeled phospholipid, *J. Am. Chem. Soc.,* 93, 314, 1971.

13. **Weber, G.,** Polarization of the fluorescence of solutions, in *Fluorescence and Phosphorescence Analysis,* Hercules, D.M., Ed., John Wiley and Sons, New York, 1966.

14. **Kinoshita, K., Jr.,** Depolarization method and molecular motion in membranes, *Nippon Butsuri Gakkaishi,* 37, 479, 1982.

15. **Chapman, D. and Dodd, G.H.,** Physicochemical probes of membrane structure, in *Structure and Function of Biological Membranes,* Rothfield, L.I., Ed., Academic Press, New York, 1971.

16. **Kito, M., Aibara, S., Kato, M., Ishinaga, M., and Hata, T.,** Effect of change in fatty acid composition of phospholipid species on the $\beta$-galactoside transport system of *Escherichia coli, Biochim. Biophys. Acta.,* 298, 69, 1973.

17. **Devaux, P. and McConnell, H.M.,** Lateral diffusion in spin labeled phosphatidylcholine multilayers, *J. Am. Chem. Soc.,* 94, 4475, 1972.

18. **Traeuble, H., Teubner, M., Wooley, P., and Eibl, H.,** Electrostatic interaction at charged lipid membranes, *Biophys. Chem.,* 4, 319, 1976.

19. **Frye, C.D. and Edidin, M.,** The rapid intermixing of cell surface antigens after formation of mouse-human heterokaryons, *J. Cell. Sci.,* 7, 319, 1970.

20. **Koppel, D.E., Axelrod, D., Schlesinger, J., Elson, E.L., and Webb, W.W.,** Dynamics of fluorescence marker concentration as a probe of mobility, *Biophys. J.,* 16, 1315, 1976.

21. **Kornberg, R.D. and McConnell, H.M.,** Inside-outside transitions of phospholipids in vesicle membranes, *Biochemistry,* 10, 1111, 1971.

22. **Rosenberg, T.,** Accumulation and active transport in biological systems, *Acta Chem. Scand.,* 2, 14, 1948.

23. **Using, H.H.,** Distinction by tracers between active transport and diffusion, *Acta Physiol. Scand.,* 19, 43, 1949.

24. **Fitts, D.D.,** *Nonequilibrium Thermodynamics: A Phenomenological Theory of Irreversible Process in Fluid System,* McGraw-Hill, New York, 1961.

25. **Schultz, S.G.,** *Basic Principle of Membrane Transport,* Cambridge University Press, Cambridge, 1980.

26. **Post, R.L.,** Reminiscence about sodium, potassium ATPase, *Ann. N.Y. Acad. Sci.,* 1974, 242, 1974.

27. **Glynn, I.M.,** The transport of sodium and potassium across cell membranes, *Sci. Basis Med. Ann. Rev.,* 1966, 217, 1966. **Whittam, R.,** Potassium movements and adenosinetriphosphate (ATP) in human red cells, *J. Physiol. (London),* 140, 479, 1958.

28. **Caldwell, P.C., Hodgkin, A.L., Keynes, R.D., and Show, T.I.,** Partial inhibition of the active transport of cations in the giant axons of *Loligo, J. Physiol.,* 152, 561, 1960.

29. **Moriyama, Y., Takano, T., and Ohkuma, S.,** Proton translocating ATPase in lysosomal membrane ghosts, *J. Biochem. (Tokyo),* 95, 995, 1984.

30. **Uchida, E., Ohsumi, Y., and Anraku, Y.,** Purification and properties of $H^+$-translocating, $Mg^{2+}$-adenosinetriphosphase from vacuolar membranes of *Saccharomyces cerevisiae, J. Biol. Chem.,* 260, 1090, 1985.

31. **Stoeckenius, W. and Bogomolni, R.A.,** Bacteriorhodopsin and related pigments of halobacteria, *Ann. Rev. Biochem.,* 52, 587, 1982.

32. **Rosen, B.P. and Kashket, E.R.,** Energetics in active transport, in *Bacterial Transport,* Rosen, B.P., Ed., Marcel Dekker, New York, 1978. **Winkler, H.H. and Wilson, T.H.,** The role of energy coupling in the transport of $\beta$-galactosides by *Escherichia coli, J. Biol. Chem.,* 241, 2200, 1966.

33. **Mitchell, P.,** Vectorial chemistry and the molecular mechanism of chemiosmotic coupling: Power transmission by proti city, *Trans. Biochem. Soc.,* 4, 399. **Harold, F.M.,** *The Vital Force: A Study of Bioenergetics,* Freeman, New York, 1986.

34. **Hirata, H., Altendorf, F.M., and Harold, F.M.,** Energy coupling in membrane vesicles of *Escherichia coli, J. Biol. Chem.,* 249, 2939, 1974.

35. **Kaback, H.R.**, Molecular biology and energetics of membrane transport, in *Biochemistry of Membrane Transport*, Semenza, G. and Carafoli, E., Eds., Springer-Verlag, Berlin, 1977.

36. **Skou, J.C.**, *Overview: The Na,K-pump in Methods in Enzymology*, Vol. 156, Fleischer, S. and Fleischer, B., Eds., Academic Press, New York, 1988.

37. **deMeis, L. and Vianna, A.L.**, Energy interconversion by the $Ca^{2+}$-dependent ATPase of the sarcoplasmic reticulum, *Ann. Rev. Biochem.*, 48, 275, 1979.

38. **Lanyi, J.K.**, Halorhodopsin: a light-driven chloride ion pump, *Ann. Rev. Biophys. Biophys. Chem.*, 15, 11, 1986.

39. **Schultz, S.G. and Curran, P.F.**, Coupled transport of sodium and organic solutes, *Physiol. Rev.*, 50, 637, 1970.

40. **Roseman, S.**, The transport of sugars across bacterial membranes, in *Biochemistry of Membrane Transport*, Semenza, G. and Carafoli, E., Eds., Springer-Verlag, Berlin, 1977.

41. **Moore, J.H. and Schechter, R.S.**, Transfer of ions against their chemical potential gradient through oil membranes, *Nature*, 222, 476, 1969.

42. **Choy, E.M., Evans, D.F., and Cussler, J.**, A selective membrane for transporting sodium ion against its concentration gradient, *J. Am. Chem. Soc.*, 96, 7085, 1974.

43. **Shinbo, T., Kurihara, K., Kobatake, Y., and Kamo, N.**, Active transport of picrate anion through organic liquid membrane, *Nature*, 270, 277, 1977.

44. **Behr, J.P. and Lehn, J.-M.**, Transport of amino acids through organic liquid membranes, *J. Am. Chem. Soc.*, 95, 6108, 1973.

45. **Broun, G., Thoma, D., and Slegny, E.**, Structured bienzymatical models formed by sequential enzymes bound into artificial supports, *J. Membr. Biol.*, 8, 313, 1972.

Chapter 3

# CHARACTERISTICS OF SYNTHETIC LIQUID MEMBRANE

**Hiroshi Tsukube**

## TABLE OF CONTENTS

# I. INTRODUCTION

There are two basic types of membranes for separation science: "liquid membranes" and "solid membranes". Liquid membranes generally offer high guest-selectivity, but solid membranes promise physical stability for practical use. For large-scale applications, much research to date has concentrated on the development of liquid membranes which were both guest-selective and mechanically stable.

The addition of a carrier to a liquid membrane system, which complexes rapidly and reversibly with the desired guest species, can dramatically improve the membrane permeability and selectivity that are significantly dependent on carrier-guest interaction. Although several transport mechanisms have been demonstrated, such a carrier-mediated transport is the most suitable for design of a specific liquid membrane. Its basic mechanism is idealized in Figure 1. The guest to be separated first reacts with the carrier to form a complex which is soluble in the membrane, but not in the adjacent solutions. After diffusion across the membrane, this complex decays in the free carrier and into the original guest species. Selective transport occurs when the carrier "selectively" binds and/or releases the guest species. Many kinds of liquid membranes have been constructed in many disciplines, such as chemical engineering, inorganic chemistry, analytical chemistry, physiology, biotechnology, and biomedical engineering. Within these disciplines, they have been applied to a wide variety of uses, such as gas separation, organic compound removal, metal ion recovery, toxic waste removal, development of selective sensing devices, enzyme reaction, and recovery of fermentation products.[1,2]

Selective membrane transport is also of central importance in many biological systems. Several types of naturally occurring carriers, so-called "ionophores", are known to specifically transport alkali and alkaline earth metal cations as well as biogenetic amines across the bio-membranes (Figure 2).[3] For example, valinomycin transports $K^+$ ion much faster than $Na^+$ ion, whereas monensin selectively transports $Na^+$ ion against its concentration gradient. Although they have a wide variety of chemical structures and molecular shapes, they all function on similar principles. They accommodate the guest species in their characteristic cavities by virtue of polar guest-binding groups and make the complexes soluble in nonpolar lipid membranes. Such striking features are significantly correlated with their elegantly organized molecular structures.[4,5] We can learn many things from bio-membrane transport when we design artificial carriers for selective liquid membrane transport.

This chapter describes the basic principle of membrane transport and fundamental characteristics of artificial liquid membranes, which cover liquid- and gas-phase separation, experimentation, and applications. The object of this chapter is to review the characteristics of the carrier-mediated liquid membrane transport processes. The reader is invited to consult other chapters of this book for more specialized problems.

# II. BASIC PRINCIPLE OF LIQUID MEMBRANE TRANSPORT

A typical liquid membrane consists of an organic liquid membrane and a carrier molecule which form a thin layer between two aqueous solutions containing different guest concentrations. The carriers alter the guest-permeability and facilitate the selective diffusion across the organic liquid membrane. This facilitated diffusion produces selective transport and separation.

The liquid membrane experiment is typically performed in a U-shaped glass tube cell as shown in Figure 3, and this membrane system is called "bulk liquid membrane".[6] A lipophilic carrier, dissolved in $CH_2Cl_2$, $CHCl_3$, or other organic media (liquid membrane) is placed in the bottom of the U-tube. Two aqueous phases (Aq. I and II), source and receiving phases, are placed in the arms of the U-tube, floating on the organic membrane

" G "
I
IG
I
" G "
I
" G "
" G "
flux
Membrane


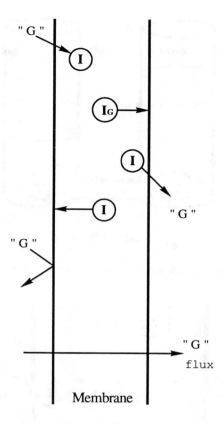

FIGURE 1. Basic mechanism of carrier-mediated membrane transport; I: carrier, G: guest.

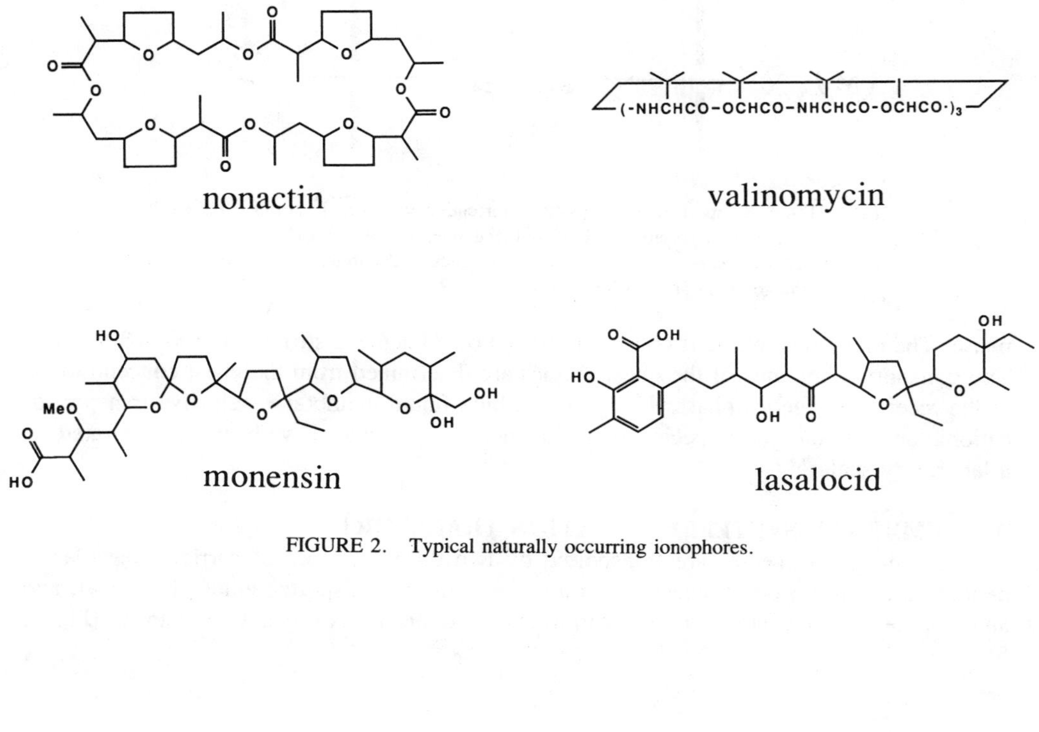

nonactin
(-NHCHCO-OCHCO-NHCHCO-OCHCO-)3
valinomycin
monensin
lasalocid


FIGURE 2. Typical naturally occurring ionophores.

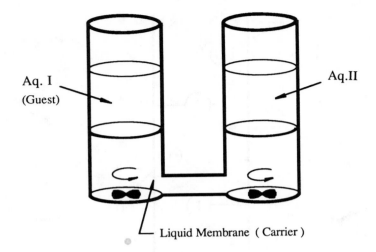

FIGURE 3. U-Shaped glass tube cell for transport experiment.

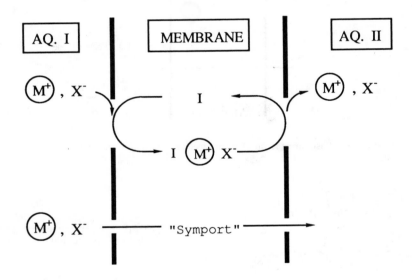

FIGURE 4. Symport membrane system for cation transport; I: carrier, $M^+$: guest cation, $X^-$: co-transported anion. (From Tsukube, H., *Analytical Applications of Functionalized Macrocyclic Host Compounds*, Industrial Publishing and Consulting, Tokyo, chapter 4, in press. With permission.)

phase. The membrane phase is constantly stirred by a magnetic stirrer (ca. 100 to 300 rpm). The transported amounts of the guest species are determined from the guest concentrations in the receiving aqueous phase. Using such a bulk liquid membrane, selective transport of cationic and anionic guest species as well as neutral molecules have been demonstrated on a laboratory scale.[7,8]

## A. MEMBRANE SYSTEM FOR CATION TRANSPORT

Cationic guest species are transported by two different types of carrier molecules: a neutral carrier transports a guest cation together with co-transported anion (Figure 4), and an anion-bearing one carries a guest cation via a cation/cation exchange mechanism (Figure 5).

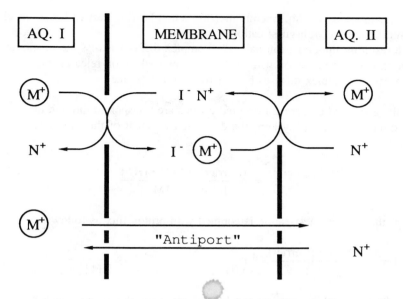

FIGURE 5. Antiport membrane system for cation transport; I⁻: anion-bearing carrier, M⁺: guest cation, N⁺: counter-transported cation. (From Tsukube, H., *Analytical Applications of Functionalized Macrocyclic Host Compounds*, Industrial Publishing and Consulting, Tokyo, chapter 4, in press. With permission.)

When a guest salt $M^+X^-$ is transported by neutral carrier such as dibenzo-18-crown-6 or naturally occurring nonactin, four elemental processes are involved ("symport"):

1. At the interface of Aq.I/membrane, guest salt is complexed with neutral carrier
2. The resulting lipophilic complex diffuses across the membrane
3. The release of guest salt occurs at the interface of membrane/Aq.II
4. The freed neutral carrier diffuses back across the membrane

In this symport, guest $M^+$ is effectively transported through the membrane together with $X^-$. The concentration gradient of the guest $M^+$ across the membrane is simply calculated from the extraction constant K of the carrier and path length L as shown below:

$$([\text{Carrier-}M^+]_{II} - [\text{Carrier-}M^+]_I)/L =$$

$$\frac{K \cdot [\text{Carrier}]_{total}}{L} \left( \frac{[M^+]_{II}}{1 + [M^+]_{II} \cdot K} - \frac{[M^+]_I}{1 + [M^+]_I \cdot K} \right)$$

Thus, the flux of the guest cation J is essentially correlated with extraction constant K and diffusion constant D:

$$J = -\frac{D \cdot K \cdot [\text{Carrier}]_{total}}{L} \left( \frac{[M^+]_{II}}{1 + [M^+]_{II} \cdot K} - \frac{[M^+]_I}{1 + [M^+]_I \cdot K} \right)$$

It is noted that the selective transport system is constructed if the employed carrier has selective extraction ability for a given guest cation.

An anion-bearing carrier such as biological monensin shows a somewhat different transport mechanism ("antiport"):

1.    At the interface of Aq.I/membrane, the anion-bearing carrier forms an electrically neutral complex with guest cation
2.    The resulting lipophilic ion-pair type complex diffuses across the membrane
3.    Cation-exchange reaction with counter-transported cation releases guest cation
4.    The carrier complex diffuses back across the membrane

As a result, guest and counter-transported cations are transported through the membrane in opposite directions. In this system, the extraction constant of the carrier K is clearly pH-dependent:

$$K = \frac{[\text{Carrier}^- \cdot M^+] \cdot [H^+]}{[\text{Carrier}] \cdot [M^+]}$$

Therefore, the flux of guest cation is coupled with proton flux as follows:

$$J = -\frac{D \cdot K \cdot [\text{Carrier}]_{\text{total}}}{L} \left( \frac{[M^+]_{\text{II}}}{[H^+] + [M^+]_{\text{II}} \cdot K} - \frac{[M^+]_{\text{I}}}{[H^+] + [M^+]_{\text{I}} \cdot K} \right)$$

Both downhill and uphill transport can be realized using this type of carrier.

## B. MEMBRANE SYSTEM FOR ANION TRANSPORT

Anion transport is similarly mediated by a carrier which possesses appropriate anion-binding sites and suitable hydrophobic portions (Figure 6). In the "antiport" transport system, a lipophilic cation-charged carrier binds the guest anion after anion-exchange at the Aq.I/membrane interface and carries it through the membrane. Then the bound guest anion is exchanged again by the counter-transported anion and released into the Aq.II phase. The net result is thus that the guest and counter-transported anions are transported in opposite directions. The "symport" transport process, in contrast, consists of different elemental processes. The guest anion and cotransported cation are complexed with a common neutral carrier, which is solublized in the membrane phase. After the complex diffuses across the membrane, it decays into the original carrier and into the original ion-pair at the other interface of the membrane. The guest anion and cotransported cation are carried in the same direction.

## C. MEMBRANE SYSTEM FOR TRANSPORT OF NEUTRAL GUEST

Neutral guests are transported between two organic phases through a water liquid membrane by water-soluble carrier molecules (Figure 7). There are many reports on cation and anion transport sytems, but few on the transport of neutral guest molecules. Microemulsion globules, ionized cyclophanes, and modified cyclodextrins were demonstrated to transport neutral guests across the water liquid membrane. $O_2$, CO, olefins, and other gaseous guests were also transported with the aid of carrier molecules through liquid membranes, but their transport mechanisms were apparently different from that shown in Figure 7.

## D. SWITCHABLE MEMBRANE TRANSPORT SYSTEM

A new and exciting area is the effect of energy input on liquid membrane transport. The use of electrical or light energy offers interesting liquid membrane systems, of which transport can be promoted or depressed by energy input.[9] For example, photochemistry is useful to provide an additional mechanism for the dissociation of the guest-carrier complex. Photodissociation reactions are common for coordination compounds and useful for acceleration of the guest-releasing process. Similarly, the electrochemistry associated with the complex can be used to release guest species, because the guest-coordination to most metal

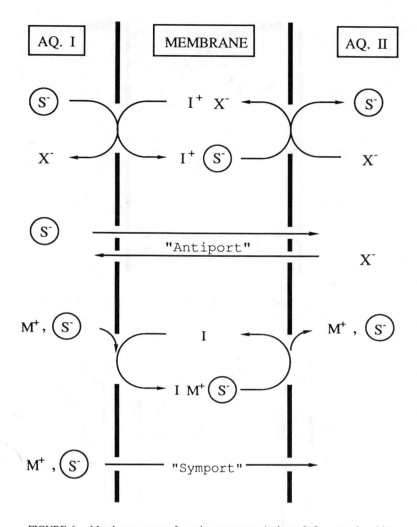

FIGURE 6. Membrane system for anion transport: Antiport & Symport; I and I$^+$: carrier, S$^-$: guest anion, X$^-$: counter-transported anion, M$^+$: co-transported cation. (From Tsukube, H., *Analytical Applications of Functionalized Macrocyclic Host Compounds*, Industrial Publishing and Consulting, Tokyo, chapter 4, in press. With permission.)

ions varies with the oxidation state of the metal center. Use of electrochemical reactions to regulate carrier-mediated transport is briefly depicted in Figure 8.

## E. MEMBRANE REACTOR

In biotechnology applications, "membrane reactors" have received great attention as they relate to the integration of the reaction, separation, and concentration of an enzymatic conversion. As schematically shown in Figure 9, a sandwich of a permselective "immobilized liquid membrane" and a "catalytic membrane" is used to allow the diffusion of the reactant into the catalytic membrane and to reject the diffusion of the product of the enzymatic reaction into the catalytic membrane. Such a membrane reactor can be applied to clean reaction and potential biomimetic sensory systems.

## III. SYNTHETIC DESIGN OF LIQUID MEMBRANE

Design of a carrier-mediated liquid membrane requires two basic steps: choice of geometry for the liquid membrane and design of the carrier molecule.

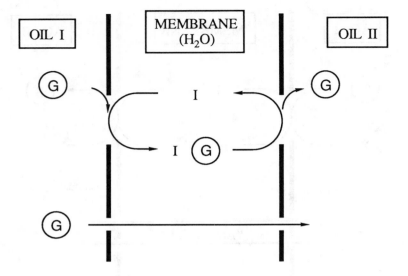

FIGURE 7.    Water liquid membrane for transport of neutral guest; I: carrier, G: neutral guest.

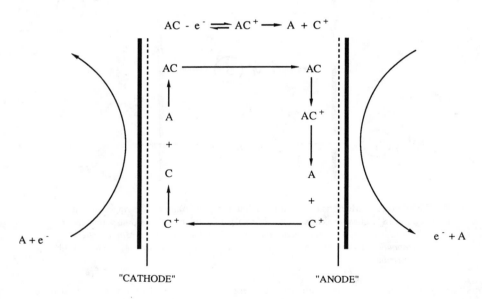

FIGURE 8.    Electrochemically switchable membrane transport system; A: guest, C: redox active carrier, e⁻: electron.

Choice of geometry for the liquid membrane depends on how the liquid membrane can be stabilized. Obviously, making the liquid membrane very thin increases its transport efficiency, but decreases the mechanical stability of the membrane. No selective transport is, of course, possible if the membrane ruptures.

Design of a carrier molecule is somewhat similar to the design of chelating reagent for a liquid-liquid extraction experiment. It is easy to screen systems for potential membrane carriers by performing liquid-liquid extraction first. Recently a more precise design of synthetic carriers has been investigated as found in Chapter 4 of this book.

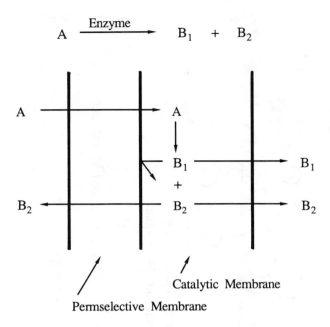

FIGURE 9.. Schematic illustration of membrane reactor; A: reactant, $B_1$ and $B_2$: product.

## A. DESIGN OF MEMBRANE MATERIAL

For practical use, "immobilized liquid membrane" and "emulsion liquid membrane" are more useful than "bulk liquid membrane" as mentioned above.[2]

"Immobilized liquid membrane" consists of a polymer support impregnated with a viscous solution of carrier (Figure 10). Many different types of polymer supports are available to prepare immobilized liquid membranes, including cellulose acetate reverse osmosis membrane, microporous polypropylene ultrafiltration membrane, polyvinylchloride filter, and hollow fiber membrane.[10] Some membranes can be cut and handled without damaging their properties. In these systems, the carrier molecules often function as mobile carriers even when they are sterically trapped within the polymer matrix.

The immobilized liquid membranes have two critical experimental problems: loss of solvent and loss or deactivation of carrier. One of the most promising approaches used to prepare stable liquid membrane is to use an ion-exchange membrane as a support. Since the carrier is retained by electrostatic force, it cannot be easily washed out of the membrane, and a useful operating life can be extended.[11]

"Emulsion liquid membranes" are easy to make, but hard to make well.[12] As shown in Figure 11, one first uses rapid stirring to make a water-in-oil emulsion, which is stabilized by an oil-soluble surfactant. When the emulsion is added, with moderate stirring, to a second aqueous phase, one obtains the desired emulsion liquid membrane. To be stable, the membrane must have a high viscosity, a neutral buoyancy, and a large concentration of surfactant. The combination of these factors should be carefully chosen.

## B. DESIGN OF CARRIER MOLECULE

Liquid membrane for practical separation purposes requires additional research on the synthesis of selective and stable carrier. It should be noted that many of the guidelines and much of the knowledge gained from the extraction systems facilitate design of powerful membrane carriers.[13] Typical examples of promising extraction reagents are summarized in Table 1.

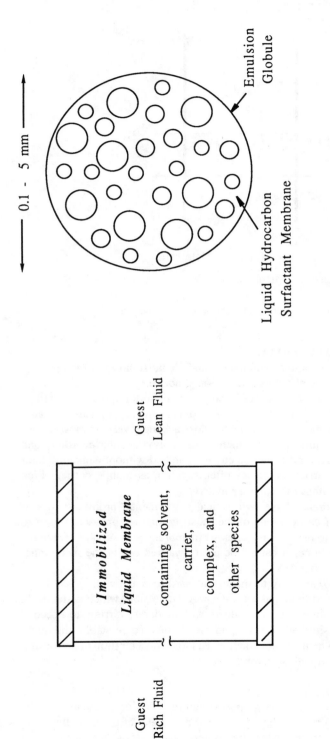

Emulsion Liquid Membrane

Immobilized Liquid Membrane

FIGURE 10.   Immobilized liquid membrane and emulsion liquid membrane.

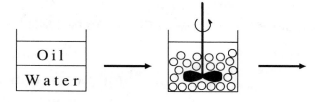

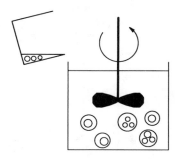

FIGURE 11. Preparation of emulsion liquid membrane. (From Tsukube, H., *Analytical Applications of Functionalized Macrocyclic Host Compounds*, Industrial Publishing and Consulting, Tokyo, chapter 4, in press. With permission.)

Numerous recent reports have concerned model studies on biomembrane transport processes.[6,8] Such data are useful and should enable the designing of better carriers of desired guest species. The general characteristics can be delineated from the naturally occurring ion-carriers (ionophores). They have several different structural features from simple chelating reagents which are elegantly organized to offer fast and selective ion-transport. For example, nonactin is the most famous naturally occurring ionophore specific for $K^+$ ion, which has a 32-membered ring. Since an 18-membered ring is known to fit the size of $K^+$ ion, the ring-size of the nonactin is apparently too large to bind a $K^+$ ion strongly. Its long chain undulates as schematically shown in Figure 12.[14] There, donor oxygen atoms form a three-dimensional cage which accommodates $K^+$ ion (1.33 Å diameter) much more effectively than small $Na^+$ ion (0.95 Å diameter). Although such encapsulated complexations are not attained using simple chelating reagents, they offer unique cation-selectivity, high stability, high dynamics, and high lipophilicity. Therefore, chemists can begin to design or search for synthetic carriers with the hearts of naturally occurring ionophores.

The rates of carrier-mediated transport generally display a bell-shaped dependence on the stability (or extraction) constants of the employed carriers. They decrease when there is either too little guest extraction at the entry or too much extraction at the exit of the membrane. In a $CHCl_3$ bulk liquid membrane system, Izatt et al. reported that the maximum transport occurred for crown-type carriers having log K values in MeOH from 5.5 to 6.0 for $K^+$ and $Rb^+$ ions and 6.5 to 7.0 for $Ba^{2+}$ and $Sr^{2+}$ ions. On the other hand, little or no transport was observed with carriers having log K less than 3.5 to 4.0.[15] The carrier must have a high solubility and stability in the liquid membrane as carrier and complex and form complexes reversibly and selectively with the guest species.

## IV. APPLICATION OF LIQUID MEMBRANE

Since the invention of liquid membranes by Robb and Li,[16,17] great progress has been

## TABLE 1
### Name and Structure of Several Types of Extractants for Metal Ion Separations[a]

| Name | Structure |
|---|---|
| "Anion exchanger": Prime, Aliquat 336 | $(CH_3)_3C(CH_2C(CH_3)_2)_4NH_2$<br><br>$R_3N(CH_3)^+Cl^-$ |
| "Acidic extractant": Di-2-ethylhexylphosphoric acid, Versatic 10, Synex 1051 | $(C_4H_9CH(C_2H_5)CH_2O)_2PO_2H$<br><br>$R_3CCO_2H$ |
| "Solvating extractant": Tributyl phosphate, Trioctylphosphine oxide | $R_3PO$<br><br>$R_3PO$ |
| "Chelating extractant": Kelex 100, LIX 65N, Polyols | |

[a]   From Tsukube, H., *Analytical Applications of Functionalized Macrocyclic Host Compounds*, Industrial Publishing and Consulting, Tokyo, chapter 4, in press, with permission.

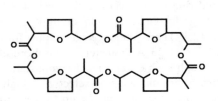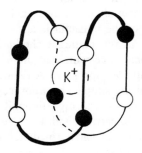

FIGURE 12.   Structures of nonactin and its K$^+$ complex.

made in the study of this novel separation technology. Several pioneering efforts are summarized in Table 2.

The earliest examples were found in carrier-mediated gas transport; for example the transport of $O_2$ using the hemoglobin-oxyhemoglobin reaction. These processes are generally categorized as facilitated transport.[18] More recently, the use of liquid membrane has been extended to the ion-transport from aqueous solutes.[19] The carrier molecule couples the flow of two or more species, and one of them can be moved against its concentration (or chemical potential) gradient. Thus, such a liquid membrane system can be considered as a "chemical pump" and applied for practical concentrations.

**TABLE 2**
**Several Pioneering Works in Liquid Membrane Science**

| | | |
|---|---|---|
| 1967 | Robb | $CO_2$ transport through liquid membrane |
| 1968 | Li | Hydrocarbon separation via O/W/O emulsion |
| 1970 | Ward | NO transport through liquid membrane |
| 1971 | Cussler | $Na^+$ uphill transport mediated by monensin |
| 1972 | Li | Phenol separation via W/O/W emulsion |
| 1977 | Baker | Uranyl concentration using hollow fiber membrane |

This section mainly discusses applications of liquid membrane systems in metal ion recovery, gas separation, waste treatment, biotechnology, and biomedical engineering. Various types of liquid membranes have been developed for separation processes, and some examples are outlined below.

## A. LIQUID MEMBRANE FOR METAL ION RECOVERY

Carrier-mediated transport through liquid membranes is currently recognized as a new technology for selective separation and concentration of toxic and valuable metal ions. Typically, $Cu^{2+}$ ion recovery systems have been formulated using emulsion liquid membranes.[20] A water-in-oil emulsion was prepared with HCl in the inner aqueous phase and classical metal-chelating reagents such as benzoylacetone as the carriers in the organic membrane phase. The rates of complex formation, agitation speed, pH value of aqueous phase, concentrations of carrier and surfactant, and ratio of the emulsion volume to the continuous phase volume were variable factors in determining $Cu^{2+}$ transport efficiency. In particular, the overall transport rates were significantly controlled by rates of complex formation and diffusion in the concentration boundary layer. Cahn et al. constructed pilot plants for $Cu^{2+}$ ion recovery using similar emulsion liquid membrane.[21] Compared to the classical solvent extraction process, they concluded that such a liquid membrane process could be up to 40% less expensive. The major cost reduction was in capital expenditures.

Systematic work has been done to investigate the use of synthetic carriers such as crown ethers and related host molecules in the liquid membranes. These carriers were usually examined in the bulk liquid membranes for the transport of mono- and divalent metal cations. For a series of alkali and alkaline earth metal cations, crown compounds showed characteristic cation transport selectivities, which are significantly controlled by the "cavity-ion size" selectivity concept.[22,23] As shown in Figure 13, 18- and 21-membered crown ethers effectively transported $K^+$ and $Rb^+$ ions, while a 15-membered one favored relatively smaller $Na^+$ ion.[15] More recently, benzo-15-crown-5 was confirmed to have potential in radioisotope separation of $^6Li^+/^7Li^+$.[24] Since this showed a good selectivity factor, other types of crown compounds may offer a new and effective carrier for practical purposes.

Acyclic carriers sometimes show more excellent transport abilities of transition metal cations than corresponding cyclic ones. Macrocyclic polyamines are known to bind very strongly to transition metal cations, but rarely release them. Since acyclic analogues show lower stability constants (usually $10^4$ to $10^5$ times),[25] they can become potential candidates for use as the carriers of transition metal cations. Tsukube et al. demonstrated that N-substituted linear polyethylenimine derivatives acted as specific carriers of transition metal cations (Figure 14). Side arm-functionalizations successfully adjusted stability constants of transition metal complexes and enhanced their solubilities in the organic liquid membrane. In a $CH_2Cl_2$ liquid membrane, urea- and thiourea-functionalized polyethylenimines specifically transported $Cu^{2+}$ ion,[26] while simple polyethylenimine transported $Zn^{2+}$ ion.[27] Urea- and thiourea-binding sites on their side arms offered moderate coordinating powers for transition metal cation and effectively promoted both complexation and decomplexation processes. Molecular design of acyclic carriers lags far behind that of cyclic carriers, but the former offers several valuable activities as useful carriers.[28,29]

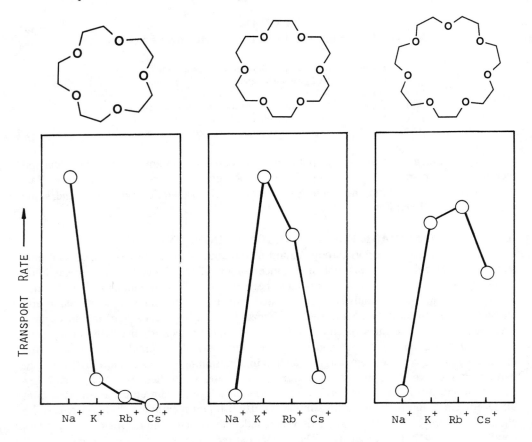

FIGURE 13.    Crown ether-mediated cation transport.

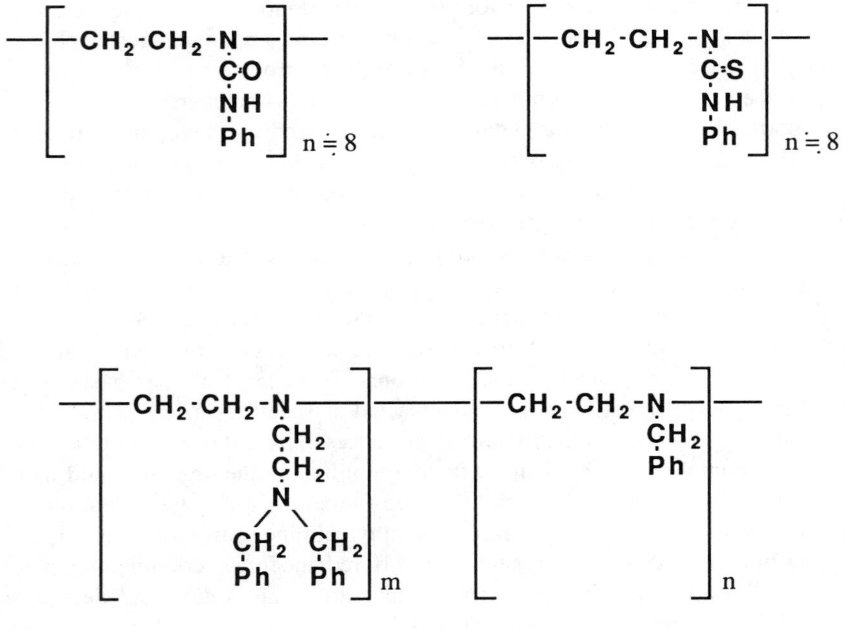

$$\left[-CH_2\text{-}CH_2\text{-}N - \begin{array}{c} | \\ \overset{..}{C}O \\ \overset{..}{N}H \\ Ph \end{array}\right]_{n \doteq 8}$$

$$\left[-CH_2\text{-}CH_2\text{-}N - \begin{array}{c} | \\ \overset{..}{C}\text{:}S \\ \overset{..}{N}H \\ Ph \end{array}\right]_{n \doteq 8}$$

$$\left[-CH_2\text{-}CH_2\text{-}N - \begin{array}{c} | \\ \overset{..}{C}H_2 \\ \overset{..}{C}H_2 \\ N \\ \end{array}\right.$$

FIGURE 14.    Acyclic carriers of transition metal cations.

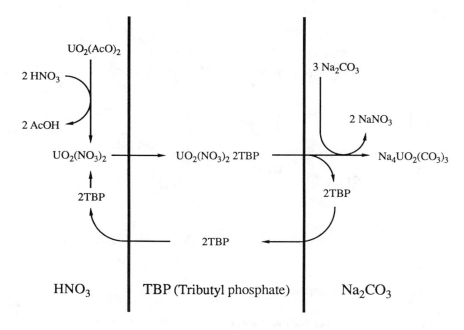

FIGURE 15.   Carrier-mediated transport of uranyl ion.

Uphill transport of uranyl ion has been extensively investigated due to its great practical importance. It was typically achieved with cellulose triacetate membrane containing tributyl phosphate as a carrier.[30] The membrane was used to separate two aqueous solutions, one acidic and the other alkaline, which were saturated with the carrier to prevent its loss from the membrane phase (Figure 15). When the initial concentration of uranyl ion in the two aqueous solutions was 3.5 m$M$, more than 50% of the uranyl ion contained in the acidic solution was trapped in the alkaline solution across the liquid membrane within 5 h. The transport mechanism involves three simple chemical reactions:

1.   Uranyl acetate dissolved in the $HNO_3$ is converted to uranyl nitrate
2.   At the $HNO_3$/membrane interface, uranyl ion is extracted by the carrier together with nitrate anion
3.   The dissociation of the uranyl nitrate-carrier complex occurs at the membrane/$Na_2CO_3$ interface

When the flow of uranyl ion is coupled to the downhill flow of nitrate anion in the same direction, the uranyl ion can be pumped uphill. Other extraction reagents such as trioctyl amine and di-2-ethylhexyl phosphoric acid, which are employed in the hydrometallurgical extraction, are also available as the carriers for concentration of uranyl ion.[31] According to an economic study based on data from a pilot plant, the capital costs of the liquid membrane separation were superior to solvent extraction when extensive feed pretreatment was required.

The precise molecular design of specific carriers for uranyl ion was elegantly performed, based on the "host-guest chemistry". Since the uranyl complexes have usually adapted either pseudoplanar penta- or hexa-coordinate structures, Tabushi et al. arranged six carboxylate anions as favored donor groups on the complementary positions of hydrocarbon rings (Figure 16).[32,33] When the ring size was adjusted to comfortably provide a cavity to accommodate uranyl ion, three of the six anionic donor groups composed a strong binding sphere for "pseudo-planar hexa-coordination". Indeed, this type of carrier offers high complexation stability and suitable transport efficiency for uranyl ion.

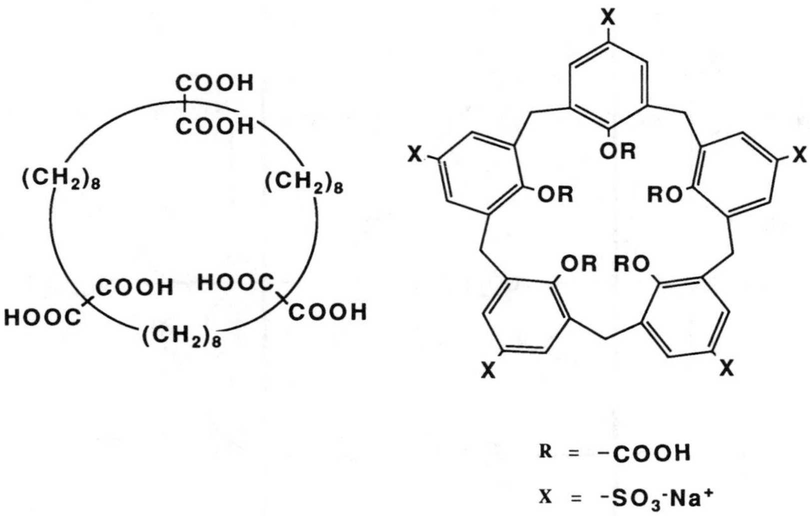

FIGURE 16.    Macrocyclic carriers of uranyl ion.

Shinkai et al. chose calixarenes as a macrocyclic backbone for synthesis of uranyl ion-specific carrier.[34] The introduction of carboxylate groups into each benzene unit of the calixarene also provided the pseudoplanar penta- or hexa-coordination structures. Their calixarene derivatives have not only high stability constants (log K = 19), but also unusually high selectivities for uranyl ion (ratio of stability constants for $UO_2^{2+}$ to $Ni^{2+}$, $Zn^{2+}$, or $Cu^{2+}$ ion: $10^{12}$ to $10^{17}$). Such molecular architecture for ion-specific carriers is a promising and effective method for modern carrier synthesis.

## B. LIQUID MEMBRANE FOR TRANSPORT OF ORGANIC CATION

Recent studies with bioscience and biotechnology relate to separation and detection of amino acid, nucleic acid, and other organic solutes. Using lipophilic anionic surfactant molecules as simple carriers, amino acid, acetylcholine, and related organic cations are transported across the liquid membrane. Typically, transport of amino acids against their concentration gradient, pumped by protonation and deprotonation, has been realized with the aid of negatively charged carriers such as dinonylnaphthalenesulfonate (Figure 17).[35] Their transport rates followed the lipophilicities of the guest cations, and tryptophane (Trp) and phenylalanine (Phe) were very effectively transported.

Sunamoto et al. presented a new class of carriers for transport of amino acid derivatives.[36] They demonstrated that a merocyanine dye would permit the transport of zwitterionic Phe across a liposomal bilayer membrane (Figure 18A). Upon UV irradiation, this is easily converted to a ring-opened form bearing zwitterion characters. Since the resulting zwitterionic carrier seemed to have complementary binding sites to zwitterionic amino acid guests, they proposed double ion-pair formation between carrier and guest ions in the membrane.

Recently Rebek et al. also developed a zwitterionic carrier for transport of amino acid derivatives (Figure 18B).[37] Its structure incorporates carboxyl groups which converge to effectively accommodate amino acid guests in a "molecular cleft". This showed an interesting transport selectivity, though its transport mechanism is somewhat unclear.

Cram and Lehn have transported the primary ammonium cations of amino acid esters, and asymmetric crown ethers have been successful in chiral recognition.[38,39] Typical examples are summarized in Table 3, including the enantiomer-selectivities in the transport and extraction experiments. In all cases, the transport selectivities were parallel to those in extraction. Thus, enantiomer of guest ammonium cation that was more rapidly transported

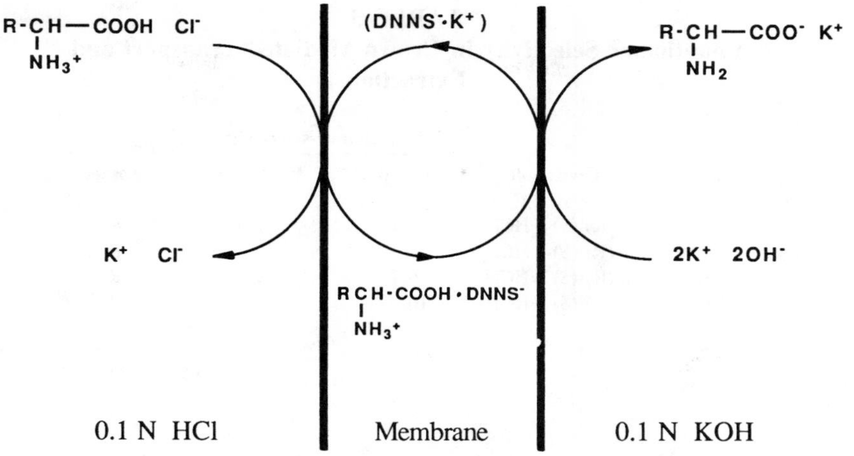

FIGURE 17. Carrier-mediated transport of amino acid cation; DNNS⁻: dinonylnaphthalenesulfonate (carrier).

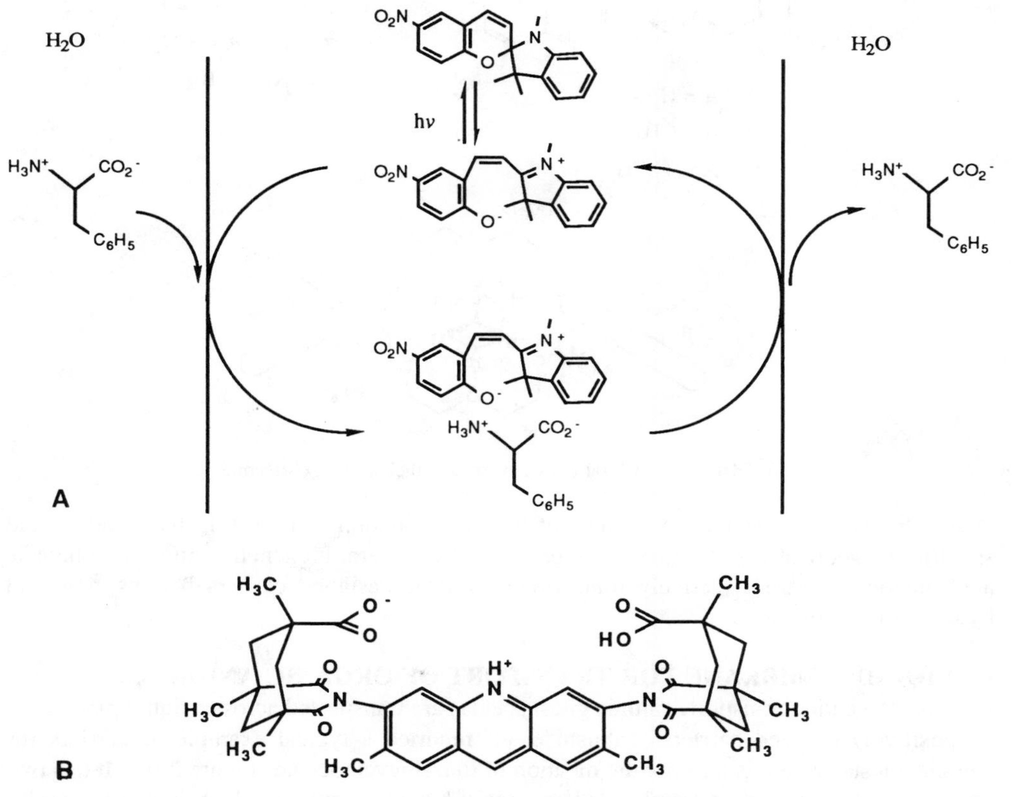

FIGURE 18. Zwitterionic carriers for amino acid transport.

was more strongly complexed via the three-point binding model as shown in Figure 19. These chiral crown ethers were recently incorporated in the supported liquid membrane and offered the possibility for practical separation and detection of racemic amino acid derivatives.[40] Detailed discussion on this is presented in Chapter 6 of this book.

Another type of synthetic carriers of organic solutes has also been developed by Tsukube

**TABLE 3**
**Enantiomer-Selectivity in Crown-Mediated Transport and Extraction**

| Crown | Guest salt | Enantiomer-selectivity | | Preferred isomer |
|-------|-----------|------------------------|------------------|-------|
| | | Transport[a] | Extraction[b] | |
| (S,S)-1 | (R) (S)-5·HCl | 2.1 | 2.5 | S |
| (R,R)-2 | (R) (S)-5·HCl | 8 | 12 | R |
| (R,R)-2 | (R) (S)-6·HCl | 6.7 | 9 | R |
| (R,R)-3 | (R) (S)-5·HCl | 10 | — | R |

[a]  $k_a^*/k_b^*$

[b]  $K_a^*/K_b^*$

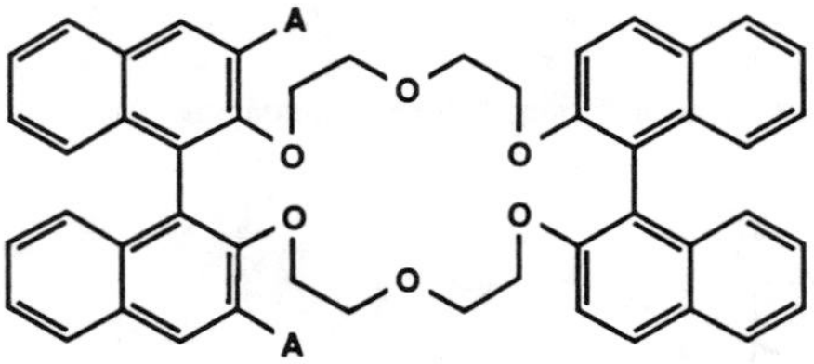

1 , A = H
2 , A = CH$_3$
3 , A = ClCH$_2$

$p\text{-}ZC_6H_4\text{-}\overset{*}{\underset{\underset{NH_3X}{|}}{C}}H\text{-}R$

4 , Z = H , R = CH$_3$
5 , Z = H , R = CO$_2$CH$_3$
6 , Z = NO$_2$ , R = CO$_2$CH$_3$

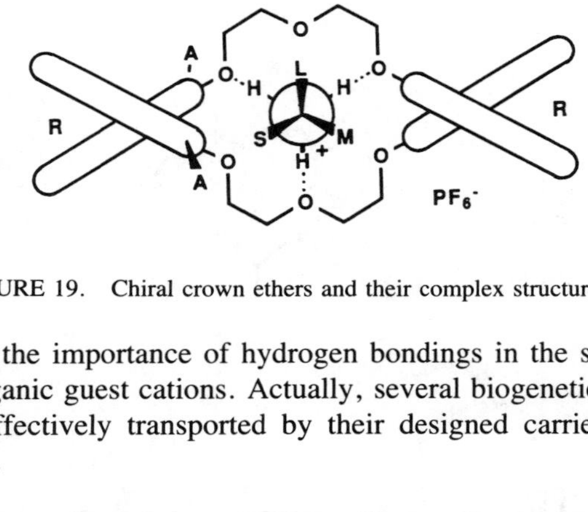

FIGURE 19.   Chiral crown ethers and their complex structure.

et al.[41,42] They noticed the importance of hydrogen bondings in the selective binding and specific transport of organic guest cations. Actually, several biogenetic amines and nucleic acid analogues were effectively transported by their designed carriers bearing powerful hydrogen bonding sites.

## C. LIQUID MEMBRANE FOR TRANSPORT OF ORGANIC ANION

Like the cation transport, anionic guest species are transported across a liquid membrane by positively charged carriers. Tabushi et al. reported a typical example of carriers for anionic guests, which is a lipophilic dication of diazabicyclooctane (Figure 20).[43] It has two cationic centers spaced and embedded in a rigid bicyclic skeleton which make it complementary to vicinal ADP dianion. Since it has less interaction with geminal AMP dianion, ADP dianion was more effectively transported through a liquid membrane. The complementary geometry between binding sites of guest and carrier was also confirmed to determine anion transport rate.

As potential anion-binding and transport carriers, macrocyclic polyammonium,[44] alkali,[45] and transition metal complexes[46] were presented. Among them, lipophilic metal complexes

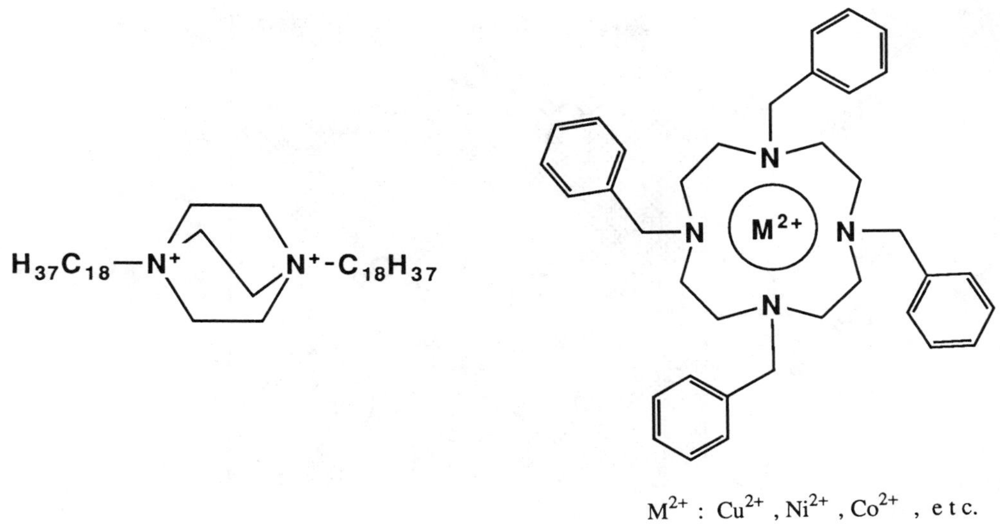

$$M^{2+} : Cu^{2+}, Ni^{2+}, Co^{2+}, etc.$$

FIGURE 20. Synthetic carriers for anion transport.

have interesting potential as a specific carrier for organic anions. As is well known, naturally occurring metallo-enzymes specifically recognize and bind the anionic guest species via "ligand protein-central metal cation-guest anion" type ternary complexations. Such metal coordination interactions may be useful in the selective binding and transport of anionic guests. Tsukube et al. were first to identify certain lipophilic transition metal complexes as a new type of anion transport carrier (Figure 20).[47] Their metallo-carriers effectively transported amino acid derivatives against their concentration gradient. Uphill transport could be mediated by these metallo-carriers.

## D. LIQUID MEMBRANE FOR GAS SEPARATION

Simple liquid membranes are often used in industrial practice for the separation of gas mixture. A typical example is the removal of $H_2S$ gas for coal gasification applications.[48] The liquid membrane consisting of an aqueous carbonate solution and a porous polymer film showed a high $H_2S$ gas permeability and great selectivity. The neutralization was proposed to occur in the transport process. Ethylenediamine was also an effective carrier for facilitated transport of $H_2S$ gas through ion-exchange membrane. It can be singly protonated to produce a carrier which can be exchanged into an ion-exchange membrane. This membrane was highly selective for $H_2S$ over $CH_4$ and had high $H_2S$ permeabilities, the seperation factors of which were up to 1200.

Transition metal complexes acted as effective carriers of various gases. In particular, the reversible binding of $O_2$ with hemoglobin and other metallo-proteins is of critical importance to advanced and primitive forms of animal life. Since the details of their $O_2$ binding and transport processes have been well characterized,[49] several synthetic metal complexes were designed as models for $O_2$ transport proteins, and some of them were useful membrane carriers for $O_2$ enrichment. Scholander and Wittlenberg were the first to study $O_2$ transport through an aqueous hemoglobin solution impregnated in their polymer membrane supports.[18] As observed in ion-transport systems, the rates of complexation and decomplexation were both important in determining overall $O_2$ transport rates. (See Figure 21)

Recently, Bend Research Inc. of Oregon developed a new liquid membrane process for $O_2$ separation from air.[50] They employed a cobalt-salen complex as a carrier and recorded $O_2/N_2$ selectivity in excess of 30. At the ambient temperature, they measured $O_2$ purity of 80%. Nishide et al. incorporated a similar cobalt-salen complex into the poly(octylmethylate-

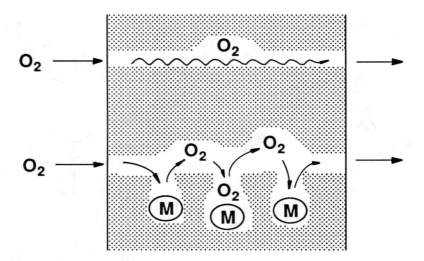

FIGURE 21.    Schematic illustration of $O_2$ transport; Ⓜ metal complex.

co-4-vinylpyridine) membrane.[51] In the polymer membrane, the cobalt complex effectively acted as a fixed carrier of $O_2$ molecule. The polymerization of a monomer containing a reactive cobalt complex was also attempted for synthesis of another type of polymer membrane.

CO separation was performed by liquid membranes containing $Cu^+$ or $Fe^{2+}$ ion. For example, $Fe^{2+}$ complex with 2,3,9,10-tetramethyl-1,3,8,11-tetraazacyclotetradeca-1,3,8,10-tetraene acted as a selective carrier of CO gas in benzonitrile-liquid membrane.[52] It showed an excellent CO selectivity over a variety of other gases. Volatile olefins such as ethylene and isoprene were also selectively carried by $Ag^+$ ion. $Ag^+$ ion-containing supported liquid membrane showed a slectivity for ethylene over ethane of approximately 1000.[53] A major problem with these metal complex carriers is that they irreversibly oxidize under the ambient conditions and require refrigeration to maintain their reversibility.

## E.  LIQUID MEMBRANE FOR ELECTRON TRANSPORT

Electron transport through a liquid membrane can be mediated by both natural and synthetic carriers which are able to undergo redox reactions. Its transport mechanisms are similar to those of anion transport systems: symport of cation and antiport of anion (see Figure 6). Typically, in biological mitochondrial and photosynthetic electron transport processes, quinone-type carriers perform the symport of two electrons and two protons via their reduced hydroquinone forms.[54]

The redox active metal complexes are suitable candidates for synthetic carriers for design of electron transport membranes. Tsukube et al. demonstrated that several lipophilic metal complexes such as bathophenanthroline complexes effectively mediated electron transport through a bulk liquid membrane.[55] Although ferrocene, bis-dithiolene, and related lipophilic metal complexes are known as effective carriers,[56] the transport rates of electron were significantly dependent on the redox potentials of carriers and the natures of co- or counter-transported species.

## F.  LIQUID MEMBRANE FOR TRANSPORT OF NEUTRAL MOLECULE

Extraction of organic solute in an aqueous solution by another immiscible solvent is usually carried out in analytical and separation processes. Liquid membranes can be similarly formulated for the removal of organic species without carrier additives. These systems rely on solubility differences between permeant guest species. Cahn and Li constructed a typical example of a liquid membrane process for phenol removal in which NaOH is encapsulated

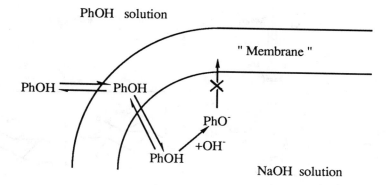

FIGURE 22.  Liquid membrane for phenol removal.

by an organic liquid membrane (Figure 22).[57] The resulting membrane phase is mixed with an outer aqueous phase containing phenol. The phenol can diffuse through the liquid membrane into the inner aqueous phase and then be neutralized to sodium phenoxide. Since phenoxide is insoluble in the organic membrane, it is trapped in the inner aqueous phase. Therefore, a high phenol concentration gradient is constantly maintained across this liquid membrane. Similar liquid membranes containing no chemical carrier were practically operated for the removal of other organic solutes such as acetic acid from the aqueous waste.

A new class of liquid membrane systems was recently constructed for transport of neutral guest species. A neutral guest molecule is carried between two organic phases through a water liquid membrane by a water-soluble carrier molecule (Figure 7). Diederich et al. prepared a water-soluble cyclophane which has a hydrophobic cavity of well-defined size and shape.[58] Since this cyclophane forms inclusion complexes with apolar pyrene, biphenyl, and durene, it acted as a unique carrier of these neutral molecules through an aqueous membrane phase. This transport is the reverse of the carrier-mediated ion transport process as metioned before and was significantly controlled by carrier-guest interactions.

Harada et al. examined methylated cyclodextrins as carriers of neutral molecules.[59] Although most inclusion complexes of cyclodextrins are insoluble both in water and in organic solvents, methylated ones effectively transported neutral azobenzene derivatives through an aqueous membrane phase. Armstrong et al. constructed a similar water liquid membrane using a series of cyclodextrins.[60] Enantiomer enrichments were interestingly demonstrated from several racemic mixtures. The transport selectivities of these systems were determined essentially by common cyclodextrin chemistry. Since other kinds of synthetic host molecules have been designed to form host-guest complexes with neutral molecules, some have potential as effective carriers providing they have suitable binding stabilities and hydrophobicities. (See Figure 23.)

## G. ION-SELECTIVE ELECTRODE

Liquid membranes are being used commercially in the production of ion-selective electrodes for a wide variety of cations and anions. Their guest-selectivities are often parallel to those observed with liquid membrane transport systems. More recently, liquid membrane systems were successfully applied in the development of gas sensors selective for organic vapors. Several interesting aspects of ion-selective electrodes will be covered in the following chapters of this book.

## V. CONCLUDING REMARKS

Several types of liquid membranes have been formulated to offer high guest-selectivity and excellent performance for metal ion recovery, gas separation, removal of organic solute,

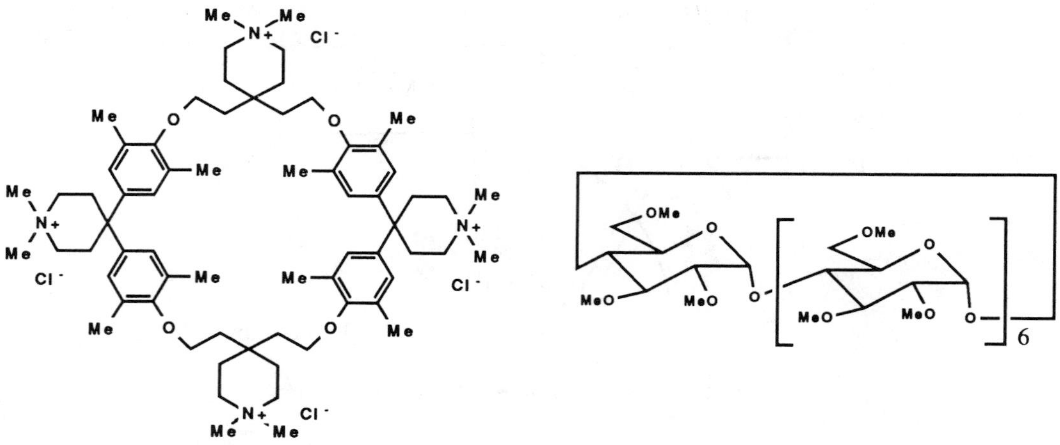

FIGURE 23.   Synthetic carriers for transport of neutral guests.

and electron transport. Since liquid membrane separation has shown itself superior in several technical and economic aspects, further fundamental and applied research promises to establish excellent practical processes based on liquid membrane science. In particular, synthetic carrier chemistry can provide new guidelines for design of carriers specific for a wide variety of guest species. Addition of carrier molecules effectively altered permeability and selectivity of liquid membrane transport phenomena.

Related theoretical work may define the set of properties required to optimize the separation and improve the existing complexation process. Recent development of membrane technology is also notable. Thin film preparation would speed industrial application of liquid membrane technology, especially in gas separation. Improved support and immobilization techniques are necessary in order to fabricate liquid membranes in thicknesses of a few micrometers or less to produce economically attractive fluxes.

In addition to the specific areas metioned in this chapter, the author believes that there are several other exciting possibilities for advancement of the utilization of liquid membrane. Investigators in the analytic, organic, inorganic, physical, medicinal, engineering, pharmaceutical chemistry, and related fields are invited to look at this new area with an eye toward innovative research and applications.

## ACKNOWLEDGMENTS

The author greatly thanks Industrial Publishing & Consulting Inc. of Tokyo, Japan, for permission to use Table 1, Figures 4 to 6, and 11. These appeared in this chapter from the author's review.[61] I also wish to acknowledge Mr. Takehiro Yoden for his assistance in manuscript preparation.

## REFERENCES

1. **Hinze, W. L. and Armstrong, D. W.,** Ordered media in chemical separations, ACS Symposium Series 342, American Chemical Society, Washington, D.C., 1987.
2. **Noble, R. D. and Way, J. D.,** Liquid membranes: Theory and applications, ACS Symposium Series 347, American Chemical Society, Washington, D.C. 1987.
3. **Painter, G. R. and Pressman, B. C.,** Dynamic aspects of ionophore mediated membrane transport, in *Topics in Current Chemistry,* 101, Springer-Verlag, Berlin, 1982, 83.

4. **Hilgenfeld, R. and Saenger, W.,** Structural chemistry of natural and synthetic ionophores and their complexes with cations, in *Topics in Current Chemistry,* 101, Springer-Verlag, Berlin, 1982, 1.

5. **Tsukube, H.,** Cation binding by natural and modified ionophores: From natural ionophore to man-made ionophore, in *Cation-Binding by Macrocyles,* Inoue, Y. and Gokel, G. W., Eds., Marcel Dekker, New York, in press.

6. **Tsukube, H.,** Biomimetic membrane transport via designed macrocyclic host molecules, *J. Coord. Chem.,* B-16, 101, 1987.

7. **Tsukube, H.,** Characteristics of new crown compounds, in *Crown Ethers and Analogous Compounds,* Hiraoka, M., Ed., Elesevier, Amsterdam, in press, chapter 3.

8. **Potvin, P. G. and Lehn, J. M.,** Design of cation and anion receptors, catalysts, and carriers, in *Synthesis of Macrocycles,* Izatt, R. M. and Christensen, J. J., Eds., Wiley-Interscience, New York, 1987, 167.

9. **Shinkai, S. and Manabe, O.,** Photocontrol of ion extraction and ion transport by photofunctional crown ethers, in *Topics in Current Chemistry,* 121, Springer-Verlag, Berlin, 1984, 67.

10. **Pusch, W. and Walch, A.,** Synthetic membranes: Preparation, structure, and application, *Angew. Chem. Int. Ed. Engl.,* 21, 660, 1982.

11. **LeBlanc, O. H., Jr., Ward, W. J., Matson, S. L., and Kimura, S. G.,** Facilitated transport in ion-exchange membranes, *J. Membr. Sci.,* 6, 339, 1980.

12. **Schwind, R. A., Gilligan, T. L., and Cussler, E. L.,** Developing the commercial potential of macrocyclic molecules, in *Synthetic Multidentate Macrocyclic Compounds,* Izatt, R. M. and Christensen, J. J., Eds., Academic Press, New York, 1978, 289.

13. **Hinze, W. L.,** Organized surfactant assemblies in separation science, in Hinze, W. L. and Armstrong, D. W., Eds., Ordered media in chemical separations, ACS Symposium Series 342, American Chemical Society, Washington, D.C., 1987.

14. **Dobler, M., Dunitz, J. D., and Kilbourn, B. T.,** Die Structur des KNCS-Komplexes von Nonactin, *Helv. Chim. Acta,* 52, 2573, 1969.

15. **Lamb, J. D., Christensen, J. J., Oscarson, J. L., Nielsen, B. L., Asay, B. W., and Izatt, R. M.,** The relationship between complex stability constants and rates of cation transport through liquid membranes by macrocyclic carriers, *J. Am. Chem. Soc.,* 102, 6820, 1980.

16. **Ward, W. J., III and Robb, W. L.,** Carbon dioxide-oxygen separation: Facilitated transport of carbon dioxide across a liquid film, *Science,* 156, 1481, 1967.

17. **Li, N. N.,** U.S.Patent 3410794, 1968.

18. **Scholander, P. F.,** Oxygen transport through hemoglobin solutions, *Science,* 131, 585, 1960.

19. **Schiffer, D. K., Choy, E. M., Evans, D. F., and Cussler, E. L.,** More membrane pumps, *AIChE J.,* 70, 150, 1974.

20. **Kondo, K., Kita, K., Koide, I., Irie, J., and Nakashio, F.,** Extraction of copper with liquid surfactant membranes containing benzoylacetone, *J. Chem. Eng. Jpn,* 12, 203, 1979.

21. **Noble, R. D. and Way, J. D.,** Applications of liquid membrane technology, in Hinze, W. L. and Armstrong, D. W., Eds., Ordered media in chemical separations, ACS Symposium Series 342, American Chemical Society, Washington, D.C., 1987.

22. **Lehn, J. M.,** Supramolecular chemistry: Scope and perspectives, *Angew. Chem. Int. Ed. Engl.,* 27, 89, 1988.

23. **Cram, D. J.,** The design of molecular hosts, guests, and their complexes, *Angew. Chem. Int. Ed. Engl.,* 27, 1009, 1988.

24. **Yashiro, O., Takano, T., Nishizawa, K., and Hiratani, K.,** private communication; also see chapter 6 of this book.

25. **Izatt, R. M., Bradshaw, J. S., Nielsen, S. A., Lamb, J. D., and Christensen, J. J.,** Thermodynamics and kinetic data for cation-macrocycle interaction, *Chem. Rev.,* 85, 271, 1985.

26. **Maruyama, K., Tsukube, H., and Araki, T.,** New membrane carrier for selective transport of metal ions, *J. Am. Chem. Soc.,* 102, 3246, 1980; Highly selective membrane transport of Cu(II) ion by synthetic linear oligomer carriers, *J. Chem. Soc. Dalton,* 1486, 1981.

27. **Tsukube, H., Kubo, Y., Toda, T., and Araki, T.,** Effective $Zn^{2+}$ ion transport mediated by branched oligoethyleneimine derivatives, *J. Polym. Sci., Polym. Lett. Ed.,* 23, 517, 1985.

28. **Kishi, N., Araki, K., and Shiraishi, S.,** The specific and uphill transport of copper(II) ion by a 6.6'-diamino-2,2'-bipyridine derivatives, *J. Chem. Soc., Chem. Commun.,* 103, 1984.

29. **Echeverria, L., Delgado, M., Gatto, V. J., Gokel, G. W., and Echegoyen, L.,** Enhanced transport of $Li^+$ through an organic model membrane by an electrochemically reduced anthraquinone podand, *J. Am. Chem. Soc.,* 108, 6825, 1986.

30. **Matsuoka, H., Aizawa, M., and Suzuki, S.,** Uphill transport of uranium across a liquid membrane, *J. Membr. Sci.,* 7, 11, 1980.

31. **Babcock, W. C., Baker, R. W., Lachapelle, E. D., and Smith, K. L.,** The mechanism of uranium transport with a tertiary amine, *J. Membr. Sci.,* 7, 71, 1980.

32. **Tabushi, I., Kobuke, Y., and Yoshizawa, A.,** Macrocyclic tridithiocarbamate as a specific uranophile, *J. Am. Chem. Soc.,* 106, 2481, 1984.

33. **Tabushi, I., Yoshizawa, A., and Mizuno, H.,** Kinetic molecular design of uranophile, *J. Am. Chem. Soc.,* 107, 4585, 1985.

34. **Shinkai, S., Koreishi, H., Ueda, K., Arimura, T., and Manabe, O.,** Molecular design of calixarene-based uranophiles which exhibit remarkably high stability and selectivity, *J. Am. Chem. Soc.,* 109, 6371, 1987.

35. **Behr, J. P. and Lehn, J. M.,** Transport of amino acids through organic liquid membranes, *J. Am. Chem. Soc.,* 95, 6108, 1973.

36. **Sunamoto, J., Iwamoto, K., Mohri, Y., and Kominato, T.,** Transport of an amino acid across liposomal bilayers as mediated by a photoresponsive carrier, *J. Am. Chem. Soc.,* 104, 5502, 1982.

37. **Rebek, J., Jr., Askey, B., Nemeth, D., and Parris, K.,** Recognition and transport of amino acids across a liquid membrane, *J. Am. Chem. Soc.,* 109, 2432, 1987.

38. **Lehn, J. M.,** Macrocyclic receptor molecules, *Pure Appl. Chem.,* 51, 979, 1979.

39. **Newcomb, M., Toner, J. L., Helgeson, R. C., and Cram, D. J.,** Chiral recognition in transport as a molecular basis for a catalytic resolving machine, *J. Am. Chem. Soc.,* 101, 4941, 1979.

40. **Yamaguchi, T., Nishimura, K., Shinbo, T., and Sugiura, M.,** Enantiomer resolution of amino acids by a polymer-supported liquid membrane containing a chiral crown ether, *Chem. Lett.,* 1549, 1985.

41. **Maruyama, K., Tsukube, H., and Araki, T.,** An artificial oligomer carrier for transport of organic substrates, *J. Chem. Soc., Chem. Commun.,* 1222, 1980.

42. **Tsukube, H.,** Specific cation transport ability of new macrocyclic polyamine compounds, *J. Chem. Soc., Chem. Commun.,* 970, 1983.

43. **Tabushi, I., Kobuke, Y., and Imuta, J.,** Carrier-mediated selective transport of nucleotides through a liquid membrane, *J. Am. Chem. Soc.,* 102, 1744, 1980.

44. **Tsukube, H.,** Lipophilic macrocyclic tetramine as specific carrier of amino acid and related anions, *Tetrahedron Lett.,* 24, 1519, 1983.

45. **Tsukube, H.,** K$^+$ ion dependent, active transport of amino acid anions by macrocyclic carriers, *Tetrahedron Lett.,* 22, 3981, 1981.

46. **Maruyama, K., Tsukube, H., and Araki, T.,** Carrier-mediated transport of amino acid and simple organic anions by lipophilic metal complexes, *J. Am. Chem. Soc.,* 104, 5197, 1982.

47. **Tsukube, H.,** Active transport of amino acid anions by a synthetic metal-complex carrier, *Angew. Chem. Int. Ed. Engl.,* 21, 304, 1982.

48. **Way, J. D. and Noble, R. D.,** Hydrogen sulfide facilitated transport in perflurosulfonic acid membrane, in Hinze, W. L. and Armstrong, D. W., Eds., Ordered media in chemical separation, ACS Symposium Series, 342, American Chemical Society, Washington, D.C., 1987.

49. **Niederhoffer, E. C., Timmons, J. H., and Martell, A. E.,** Thermodynamics of oxygen binding in natural and synthetic dioxygen complexes, *Chem. Rev.,* 84, 137, 1984.

50. **Roman, I. C. and Baker, R. W.,** Japan Patent 59-12707; **Maton, S. L., and Londsdale, H. K.,** Liquid membranes for the production of oxygen enriched air. III, *J. Membr. Sci.,* 31, 69, 1987.

51. **Nishide, H., Kuwahara, M., Ohyanagi, M., Funada, Y., Kawakami, H., and Tsuchida, E.,** Oxygen permeation in the membrane of poly(octylmethacrylate-co-4-vinylpyridine)-salicylaldehydeethylenediimine cobalt complex, *Chem. Lett.,* 43, 1986.

52. **Koval, C. A., Noble, R. D., Way, J. D., Louie, B., Reyes, A., Horn, G., and Reed, D.,** Selective transport of gaseous CO through liquid membrane using a iron(III) macrocycle complex, *Inorg. Chem.,* 24, 1147, 1985.

53. **Teramoto, M., Matsuyama, H., Yamashiro, T., and Katayama, Y.,** Separation of ethylene from ethane by supported liquid membranes containing silver nitrate as a carrier, *J. Chem. Eng. Jpn,* 19, 419, 1986.

54. **Amesz, J.,** The function of plastoquinone in photosysnthetic electron transport, *Biochim. Biophys. Acta,* 301, 35, 1973.

55. **Maruyama, K. and Tsukube, H.,** Metal complex mediated electron transport system, *Chem. Lett.,* 1133, 1981.

56. **Grimaldi, J. J. and Lehn, J. M.,** Multicarrier transport: Coupled transport of electron and metal cations mediated by an electron carrier and a selective cation carrier, *J. Am. Chem. Soc.,* 101, 1333, 1979.

57. **Li, N. N.,** Permeation through liquid surfactant membranes, *AIChE J.,* 17, 459, 1971.

58. **Diederich, F. and Dick, K.,** A new water-soluble macrocyclic host of the cyclophane type: Host-guest complexation with aromatic guests in aqueous solution and acceleration of the transport of arenes through an aqueous phase, *J. Am. Chem. Soc.,* 106, 3024, 1984.

59. **Harada, A. and Takahashi, S.,** Transport of neutral azobenzene derivatives by methylated cyclodextrin, *J. Chem. Soc. Chem. Commun.,* 527, 1987.

60. **Armstrong, D. W. and Jin, L. H.,** Enrichment of enatiomers and other isomers with aqueous liquid membranes containing cyclodextrin carriers, *Anal. Chem.,* 59, 2237, 1987.

61. **Tsukube, H.,** Membrane separation of inorganic ions mediated by functionalized macrocycles, in *Analytical Applications of Functionalized Macrocyclic Host Compounds,* Takeda, H., Ed., Industrial Publishing & Consulting, Tokyo, in press, chapter 4.

Chapter 4

# SYNTHETIC STRATEGIES FOR SELECTIVE LIQUID MEMBRANES: CARRIER CHEMISTRY

## Hiroshi Tsukube

## TABLE OF CONTENTS

# I. INTRODUCTION

Liquid membranes incorporating potential chemical carriers can effect rapid and selective transport of a variety of guest species and provide new possibilities in membrane separation. They can concentrate guest solutes of interest, converting a dilute mixture into a concentrated solution of one desired component. Further development of such liquid membranes requires a sophisticated combination of carrier chemistry and membrane science. This chapter describes recent developments in carrier chemistry, which provide the molecular design of powerful carrier molecules specific for various guest species. We mainly discuss the guest-binding and transport functions, both of synthetic and of biological carriers in the bulk liquid membrane systems. Their target guest species include alkali, alkaline earth, heavy and transition metal cations as well as organic substrates. Transport mechanisms and other characteristics of carrier-mediated liquid membranes have already been detailed in Chapter 3 of this book.

# II. MOLECULAR DESIGN OF SYNTHETIC CARRIER

The selective liquid membrane system can be designed through molecular architecture of specific chemical carriers. One of the most promising carrier syntheses is to mimic the biological ion-carriers, so-called "ionophores". We know many kinds of naturally occurring ionophores involved in biomembrane systems. They vary widely in chemical composition and molecular size, but all mediate efficient biomembrane transport based on a simple principle.[1,2] Thus, they offer excellent models for the molecular design of specific synthetic carriers.

Typically, biological monensin has several unique molecular structures which are common in other naturally occurring ionophores. Although it is an acyclic polyether antibiotic, it forms highly stable and lipophilic complexes with several metal cations. Such specific complexations come from the characteristic pseudo-cyclic conformation,[3] in which suitably placed $-CO_2^-$ and $-OH$ groups at the chain-ends are linked by intramolecular "head to tail" hydrogen bonds (see Figure 1). The free, neutral monensin also exists in a pseudo-cyclic conformation, but there are important differences in the relative positions of the ether-oxygen atoms between neutral and metal-complexed anionic monensins. Some of the O–O distances change by more than 1 Å on passing from $Ag^+$ salt to free acid, and monensin is "loosely" preorganized for effective accommodation of guest cation.[4] Furthermore, the monensin shows pH-dependent complexation/decomplexation switching properties, allowing highly dynamic complexations. The stability constant of $Na^+$ complex of neutral monensin is almost $10^5$ times lower than that of the anion-charged form in ethanol solution.[5] This probably results from coupling of the complexation with acid dissociation of monensin carboxylic acid. Crystallographic studies of complexes with naturally occurring ionophores clarified similar common interesting features:

1. They specifically wrap around the guest cations in a three-dimensional sense
2. They contain suitable hydrophilic and lipophilic portions, so that they can bind polar guest species and solubilize them into nonpolar membrane
3. Both complexation and decomplexation processes are highly dynamic, promising effective transport

We must search for synthetic carriers with these features.

When we design synthetic carrier molecules showing specific ion transport abilities, crown compounds are, without question, one of the most promising candidates.[6] In 1973, Reusch et al. were first to formulate a bulk liquid membrane transport, in which dibenzo-

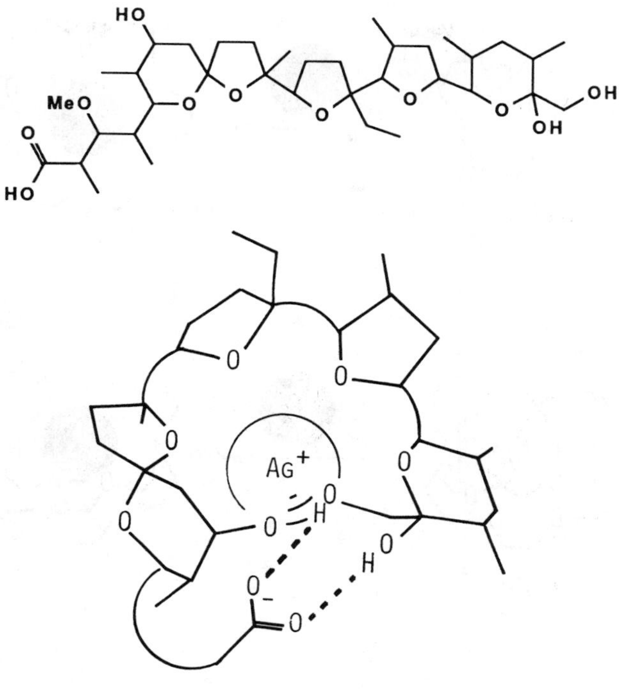

FIGURE 1.   Structures of monensin and its Ag+ complex.

18-crown-6 selectively transported alkali and alkaline earth metal cations.[7] It offered sur-prisingly high transport selectivity for $K^+$ cation ($K^+/Na^+ > 10$, $K^+/Cs^+ > 100$, and $K^+/Li^+ > 1000$). The striking features of such a simple crown ether carrier have served to extensively advance the molecular design of crown-based synthetic carriers, and have pro-vided much insight on biological membrane transport mechanism as well as chemical ap-plications in various fields. Since synthetic crown compounds have the advantages of facile synthesis and of versatile molecular structures, we can design various crown-based carriers through molecular architecture.

Many kinds of synthetic carriers are known to act primarily as the effective carriers for cationic guests.[8,9] Their molecular design has usually been done by considering

1.   Size and shape of carrier cavity for guest accommodation
2.   Nature of guest-binding site
3.   Three-dimensional coordination topology
4.   Solubility and stability as carrier and complex
5.   Molecular flexibility and binding dynamics

Although naturally occurring ionophores transport only limited kinds of biological guests such as $Na^+$ and $K^+$ cations and catecholamines across the biomembrane, synthetic carriers can mediate artificial transport of alkali, alkaline earth, heavy and transition metal cations as well as organic cations and anions. Typical examples covering some recent topics in this field are presented below.

For design of a specific host molecule, several useful synthetic strategies have been established. Cram presented the principle of "preorganization" and developed a series of extremely powerful and selective host molecules of alkali metal cations: he said, "The more highly hosts and guests are organized for effective binding and low solvation prior to complexation, the more stable are their complexes."[10] Spherand is the most completely

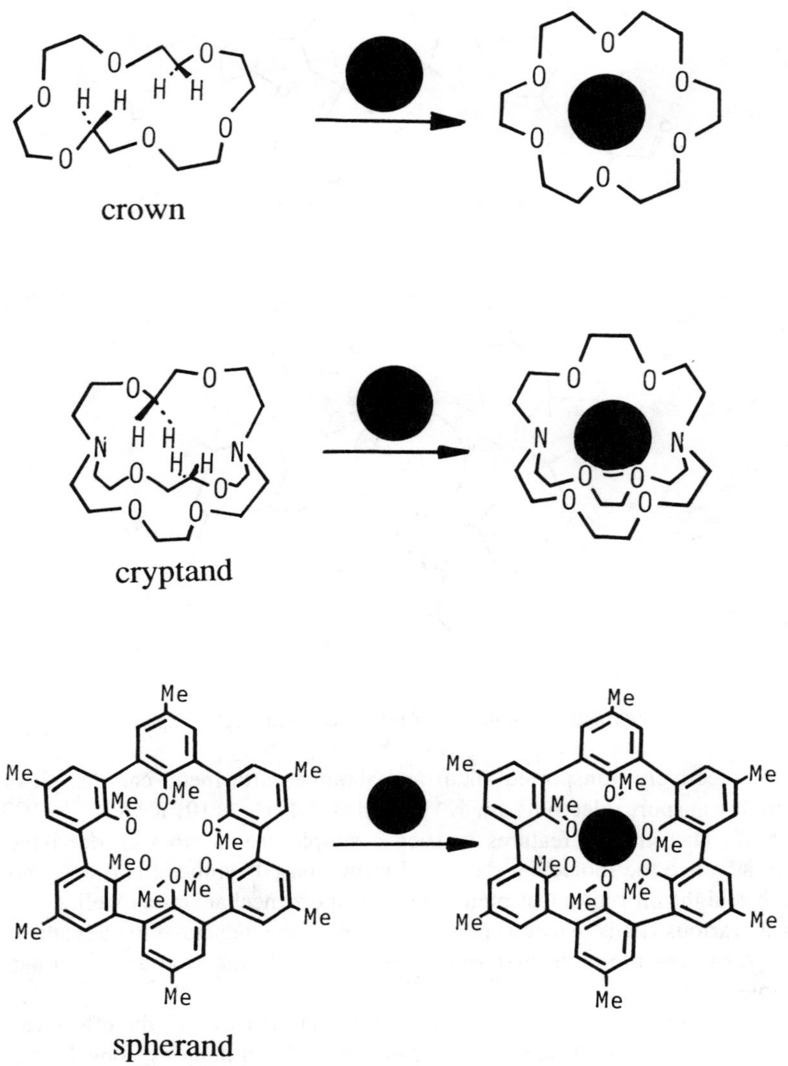

FIGURE 2. Crystal structures of crown, cryptand, spherand, and their complexes with the size-fitting guest cations.

preorganized molecular system along this line. Its crystal structure is markedly different from those of common crown ether and cryptand (Figure 2). Free crown ether and cryptand fill their own cavities with inward-turned methylene groups in crystal structures. Thus, the cavities are undoubtedly unformed, either being filled with their own methylenes or with parts of solvent molecules in solutions. Since they are displaced by a guest upon complexation, the guest cation organizes the cavity shape of host molecule. In contrast, spherand is fully preorganized for complexation, and its binding ether oxygens are completely nonsolvated both in solid and solution states. This class of molecules therefore can promise highly stable and selective complexation with a given size-fitting guest. Although we can design specific synthetic host molecules for a desired guest on the basis of this "preorganization concept", such rigid host molecules cannot be recommended as the effective carriers in dynamic transport processes.

Other synthetic strategies offer rational molecular designs for synthetic carrier mole-

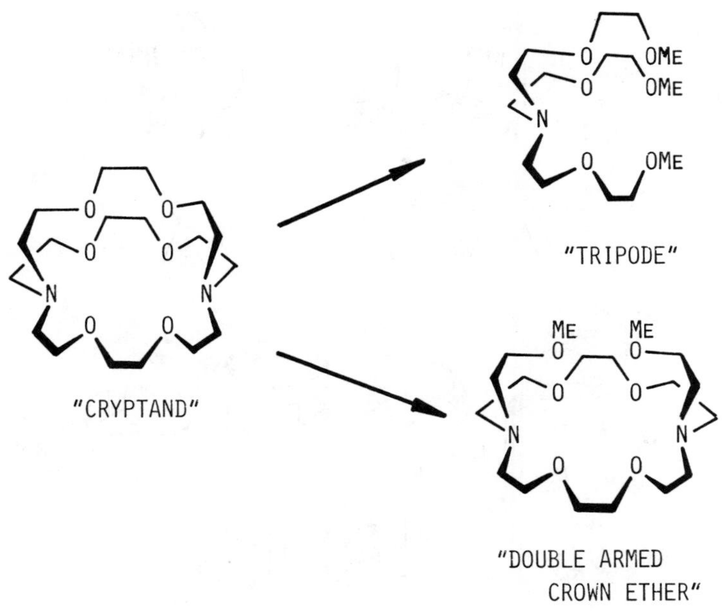

FIGURE 3.   Molecular structures of tripode and double armed crown ether.

cules.[11-14] Tsukube et al. developed the most useful approach for synthesis of potential carrier
(Figure 3). They prepared two different types of a new class of flexible host molecules via
ring-opening of bicyclic cryptand: tripode[15] and double armed crown ether.[16] These "armed
host molecules" have flexible open-chain structures, but envelop the guest cations completely
in a three-dimensional sense. Their binding dynamics are high enough to satisfactorily meet
the level required of ion-carriers. One can believe, "The more flexible host molecules have,
the higher should be the rates for complexation and decomplexation."

Since the guest-binding selectivities of these synthetic host molecules are arbitrarily
adjusted by structural modifications, we can design excellent crown-based flexible carriers
on a molecular basis.[17] The synthetic strategies and chemical functions of new synthetic
carriers are discussed below from the viewpoint of "carrier chemistry", together with
naturally occurring ionophores and their chemically modified derivatives.

## III. NEW SYNTHETIC CARRIER FOR CATION TRANSPORT

A large number of crown-based host molecules have been explored as specific binders
and extractants for cations, anions, and neutral guest species. They accommodate the guest
species in their internal cavities by virtue of ligating donor groups. Depending on their
ligand topology, they are traditionally classified into the three categories (Figure 4): acyclic
"podands" (type A), monocyclic "crown ethers" (type B), and bicyclic "cryptands" (type
C). In addition, several other new types of crown-based compounds such as armed macrocycles (type D, E, G, H, I) were also prepared, which topologically lay at the borderlines
between these three. Among them, crown ethers, armed macrocycles, and acyclic podands
provide interesting possibilities in design of synthetic carriers exhibiting excellent functions.

### A. CROWN COMPOUND

Among various crown ether families with different numbers, distributions, and types of
donor atoms,[18] we can choose a suitable combination of binding site and ring size and prepare
a chemical carrier specific for a target guest. Typically, a new series of crown-type carriers

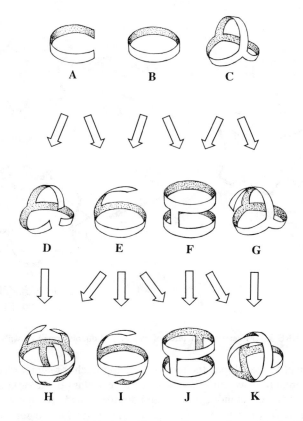

FIGURE 4.    Topological classification of crown ether and its family.

for organic ammonium cations was synthesized following a biological transport system.[19] Biological carrier proteins are known to bind and transport many organic guest species via several specific interactions such as charge-dipole attractions, hydrogen bondings, and hydrophobic interactions.[20]

Lehn et al. reported solution complexation behaviors between primary ammonium cations and synthetic N, O-mixed 18-crown-6 derivatives (Figure 5).[21] The hydrogen bonding between crown ring nitrogen atoms and guest ammonium cation was significantly involved in host-guest complexation. Among the 18-crown-6 derivatives, $N_3$, $O_3$-mixed one was found to bind primary ammonium cations via three tight hydrogen bonds most strongly. These results provide a useful basis for carrier design. Tsukube et al. prepared a new class of synthetic carriers by introducing amino and amide binding sites into the crown-ring structure and formulated a new "symport" liquid membrane for cation transport.[19,22] As shown in Figure 6, the guest cation first reacts with the neutral carrier to form a complex which is soluble in the membrane, but not in the adjacent aqueous solutions. After diffusion across the membrane, the complex decays into the original carrier and into the guest species. Selective transport results when the complexation or decomplexation is selective.

Amino- and amide-functionalized crown compounds effectively transported several organic ammonium cations, whereas $Na^+$, $K^+$, and $NH_4^+$ ions were rarely transported under the same conditions. Dibenzo-18-crown-6 showed high transport efficiencies both for organic ammonium and $K^+$ cations of similar sizes, but hardly discriminated between them. Detailed transport selectivities for a series of amino acid ester salts were interestingly adjusted by choosing donor-sites in macrocyclic carriers. Amino-crown compounds transported glycine (Gly) and alanine (Ala) ester salts more effectively than phenylalanine (Phe) and tryptophane

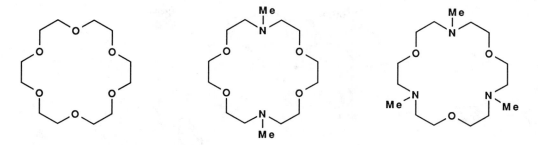

FIGURE 5. N,O-Mixed 18-crown-6 derivatives for binding of primary ammonium cations.

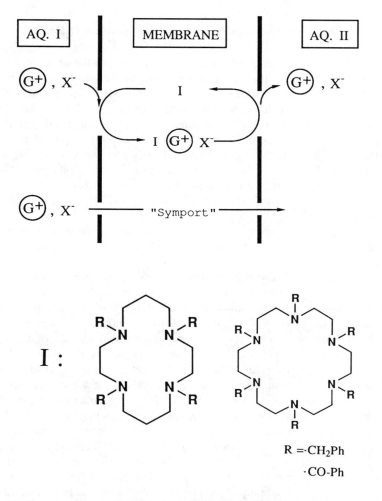

FIGURE 6. Symport liquid membrane and neutral carriers for transport of organic cations: I: carrier, $G^+$: guest cation, $X^-$: co-transported anion.

(Trp) ester salts, though amide-ones showed higher transport rates for Phe and Trp derivatives. These two carrier systems probably had different rate-determining steps: the releasing process of guest cation determined the overall transport rate of the polyamine carrier system, and the extraction process essentially governed the polyamide-mediated transport system. Thus, both transport efficiency and selectivity were controlled by altering donor site in crown-based carrier.

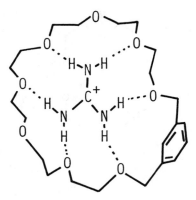

FIGURE 7.    Crystal structure of benzo-27-crown-8 — guanidinium cation complex.

Large-membered crown compounds acted as unique carriers for large-sized guest cations. The size of crown ring generally determines the size of the guest cation to be complexed and transported: 18-membered crowns are selective for $NH_4^+$ salt whereas 27-membered ones prefer guanidinium salt. X-ray crystallography indicated that 27-membered crown ether had a suitable cavity to form six hydrogen bondings with polyfunctional guanidinium cation (Figure 7). Indeed, the guanidinium cation was effectively transported by crown ethers having more than 27-membered rings.[23] Their transport rates across a bulk liquid membrane were correlated well to their complex stability constants. 27- and 30-membered benzo-crown ethers offered faster transport of guanidinium thiocyanate than 18- and 33-membered ones. Imidazolium and other biological organic cations are the next targets of these large-membered crown carriers.

Anion-charged crown compounds efficiently mediated cation transport via antiport mechanism (Figure 8). The guest cation forms an electrically neutral complex with the anionic carrier at the one membrane side, and is exchanged by the appropriate cation (or proton) at the other membrane side. Although anionic carriers easily form intermolecular ion-pair complexes under alkaline conditions, some strong acids are usually required for fast transport. Izatt et al. examined p-tert-butylcalixarenes as the carriers which have proton-dissociative phenol moities.[24] If the aqueous phase I was an alkaline solution of metal hydroxide, effective cation transport occurred. Since they could not form encapsulated metal complexes like crown compounds, the transport mediated by these anion-charged carriers was selective for less hydrophilic $Cs^+$ ion over $Li^+$, $Na^+$, and $K^+$ cations. Kimura et al. reported that a lipophilic dioxocyclam derivative mediated a proton driven $Cu^{2+}$ ion transport.[25] This macrocycle selectively bound $Cu^{2+}$, $Ni^{2+}$, $Co^{2+}$, or $Pt^{2+}$ ions under neutral or alkaline conditions, with a simultaneous deprotonation of the two amides to yield a stable 1:1 complex. Since the resulting complexes underwent immediate dissociation when exposed to strong acid, transition metal cations were effectively transported against their concentration gradients.

Redox-active crown compounds were recently demonstrated to show interesting cation-binding properties which were coupled to the electrochemical reactions. Saji et al. constructed an electrochemically switchable liquid membrane system in which ferrrocene-functionalized crown compound transported guest cation and electron in the same direction (Figure 9).[26] At the left interface of the membrane, ferrocene-functionalized crown compound acted as a neutral carrier and accommodated a $Na^+$ ion in a crown ring. The complex diffused to the right interface, whereupon it became oxidized at the electrode $W^1$. The oxidized crown might lose the guest cation to the aqueous phase via electric repulsion between ferrocenium and guest cations. After diffusion to the left interface, it became reduced at electrode $W^2$ to recommence the cycle. The electrochemical energy can be utilized as the driving force in the cation transport[27,28] as well as pH gradient[29] and light energies.[30,31] More detailed discussion on such switchable liquid membrane transport will be presented in Chapter 7.

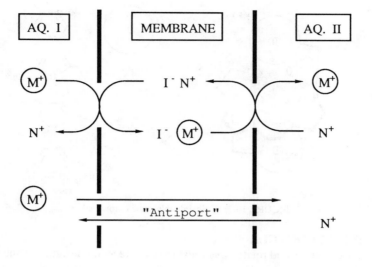

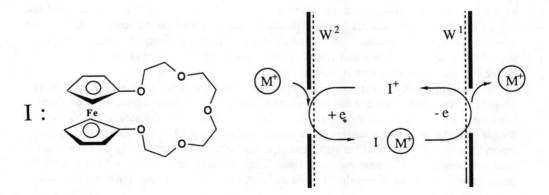

FIGURE 8.  Antiport liquid membrane and anion-charged carriers for transport of cations; I⁻: carrier, M⁺: guest cation, N⁺: counter-transported cation.

FIGURE 9.  Ferrocene-functionalized crown ether and its membrane transport system; M⁺: guest cation, W¹ and W²: electrode, e: electron.

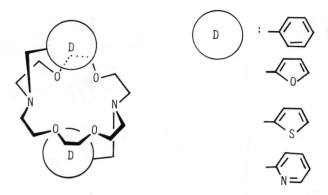

FIGURE 10.    Typical double-armed crown ethers.

## B.  ARMED MACROCYCLE

Armed crown ethers and related macrocycles were recently designed as the most suitable carriers for cation transport.[11] They form three-dimensional and lipophilic complexes with various guest cations, while the high mobility of ligating side arms attached to the parent macro-ring promises highly dynamic complexation.[32] Therefore, we can select a combination of the parent macrocyclic structure and ligating side arm suitable for the given guest cation to be transported.

Tsukube et al. first presented systematic studies on cation transport properties of double-armed crown compounds (Figure 10).[16,33] The replacement of benzene ring in the N,N'-dibenzyldiaza-18-crown-6 by cation-ligating furan (oxygen atom), thiophene (sulfur atom), and pyridine (nitrogen atom) groups largely increased transport efficiencies for complementary guest cations. Among a series of diaza-18-crown-6 derivatives, furan-armed crown compound mediated transport of $K^+$, $Ba^{2+}$, $NH_4^+$, and $Pb^{2+}$ ions more effectively than simple crown compound, though other alkali and alkaline earth cations were only slightly transported. In contrast, thiophene- and pyridine-armed crown compounds offered enhanced transport rates for heavy and transition metal cations.[34]

Figure 11 illustrates relationship between cation transport selectivity and ring-size of the parent crown ring in furan-armed diaza-crown ether systems.[14] The furan-armed crown ether with a 15-membered ring effectively mediates $Na^+$ ion transport, while a 21-membered one is a good carrier of $K^+$ and $Cs^+$ ions. Therefore, double armed crown compounds have the great advantages of "tunable" guest-binding and transport selectivity. By considering ring-size of the parent crown ring and coordinating character of donor arm groups, we can easily image the molecular structure of specific carrier for a target guest cation.

In carrier-mediated transport, the transport rates always display a bell-shaped dependence on stability constants between carrier and guest species. The relationships between stability constants and transport rates of $Na^+$, $K^+$, and $Pb^{2+}$ ions are illustrated for crown ether, double armed crown ether, and cryptand molecules in Figure 12.[35] The double armed crown ethers clearly have intermediate stability constants between corresponding crown and cryptand compounds and highest transport rates. Their flexible structures can wrap around the guest cations in a three-dimensional sense and exhibit suitable binding dynamics to promote cation transport.

Attachment of an ionizable arm-group to the crown skeleton is a useful method for development of better carrier. Kimura et al. reported that phenoxide anion-armed 14-crown-4 derivatives were excellent $Li^+$ ion-specific carriers in a liquid membrane transport as well as an ion-selective electrode system (Figure 13).[36] They recorded an extremely high selectivity ratio of $Li^+$ over $Na^+$ ions. In such an anion-armed crown compound, the combination of an ion-binding crown cavity with anion-charged donor arm groups creates a powerful bifunctional complexing agent.

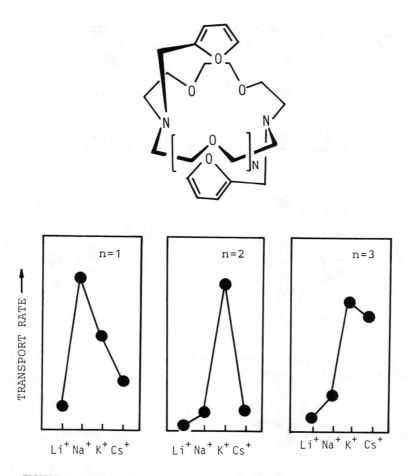

FIGURE 11. Relationship between cation transport rate and ring-size of furan-armed diaza-crown ether.

Nakatsuji et al. prepared a crown ether carrier having a positively charged arm group which mediated active transport of $K^+$ ion.[37] Since the $-NH_2$ group-bearing 18-crown-6 compound can form an intramolecular complex with $-NH_3^+$ moiety in acidic solution (see Figure 13), it showed a relatively higher complexing ability for $K^+$ ion under basic conditions than under acidic conditions. Similar reversible "tail-biting" complexation of crown ring and $-NH_3^+$ cation on the side arm prevailed in the photo-responsive crown compounds.[38]

Multi-armed aza- and thia-macrocycles form lipophilic and stable complexes with a different series of guest species from armed crown ether compounds. They can use a varying number of ligating side arms and accommodate their molecular cavities to the size and shape of the guest cation. Since ring-size and donor-atom of parent macrocyclic ligand, number and nature of arm donor group, and shape of molecular cavity are variable factors, new and specific carriers of this type can be designed at will.

Tsukube et al. demonstrated that tetraaza-macrocycles bearing amide- and pyridine-functionalized side arms had unique transport abilities for $Ba^{2+}$ and $Na^+$ ions (Figure 14).[39] Although parent 1,4,8,11-tetraazacyclotetradecane ring favored proton and some transition metal cations, arm-functionalization significantly modified carrier functions. [13]C-NMR binding studies strongly suggested that amide-armed azamacrocycle formed a stable 1:1 complex with $Ba^{2+}$ ion, in which the guest cation was located on the polyamine ring and effectively coordinated both by amide and macroring nitrogen groups. Pyridine-armed azamacrocycle

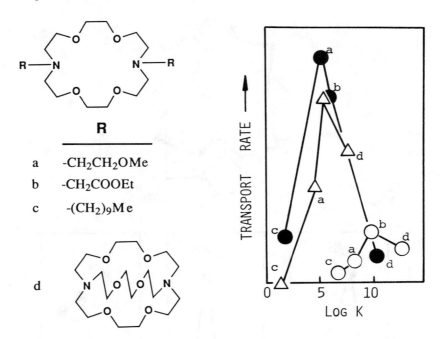

FIGURE 12.    Relationship between transport rate and stability constant in double armed crowned ether and related carriers; ●: $K^+$, △: $Na^+$, ○: $Pb^{2+}$.

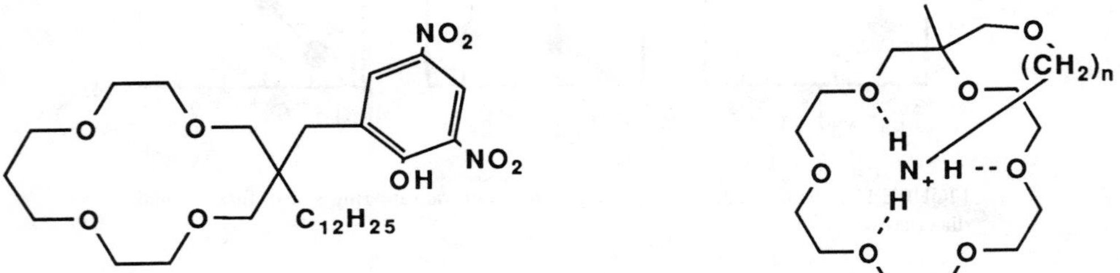

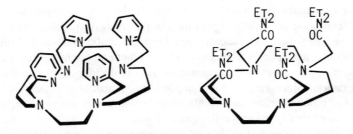

FIGURE 13.    Ionizable armed crown ether carriers.

FIGURE 14.    Multi-armed aza-macrocycles for transport of cations.

accommodated $Na^+$ ion selectively in its molecular cavity. Since such selective binding and transport of $Ba^{2+}$ and $Na^+$ ions over $K^+$ ion are not attained by common crown ether carriers, new ligand combination and geometry offer new carrier function.

Armed thia-macrocycle specifically transported $Ag^+$ ion across a $CHCl_3$ membrane, while $Na^+$, $K^+$, and $NH_4^+$ cations of similar ion sizes were rarely transported (Figure 15).[40]

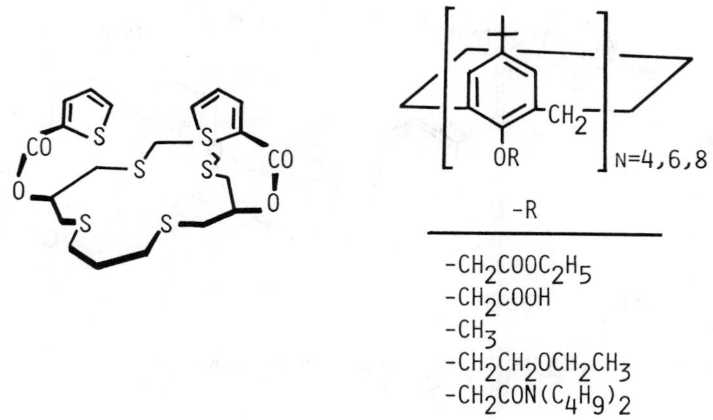

FIGURE 15.  Armed macrocycle carriers of cationic guests.

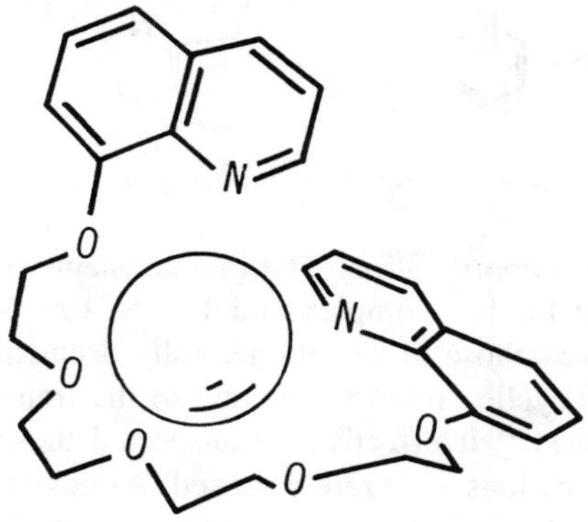

FIGURE 16.  Crystal structure of Rb$^+$ complex with quinoline-bearing acyclic crown compound.

Although Ag$^+$ ion-selective transport is a characteristic of the parent thia-crown ring, thiophene-functionalization of side arm greatly increased transport efficiency.

Chang and Cho employed calixarenes as the molecular backbones to construct a new series of armed macrocyclic carriers (Figure 15).[41] Their moderately rigid skeletons allow satisfactory high guest selectivity, while their conformational freedom is retained to assure high binding dynamics. A series of armed calixarenes was examined as the carrier for cation transport, which had various ring sizes and neutral arm donor groups. Among them, ethoxycarbonylmethyl derivatives showed the most notable transport selectivities: calix-4-arene for Na$^+$ ion, calix-6-arene for Cs$^+$ ion, and calix-8-arene for K$^+$ ion. The participation of the ester-arms may lead to encapsulated complexation and effective transportation. Kimura et al. recently utilized similar armed calixarenes as a liquid membrane-type ion-selective electrode.[42]

## C. ACYCLIC CROWN COMPOUND: PODAND

Several functionalized acyclic crown compounds can act as carriers if they form three-dimensional complexes. Quinoline-bearing acyclic crown compound forms a characteristic podand-metal complex. X-ray investigation on its Rb$^+$ complex demonstrated that acyclic polyether chain wrapped around a fitting guest cation in a helical fashion as shown in Figure 16.[43] A screw-shaped arrangement of molecular chain gained the participation of all oxygen

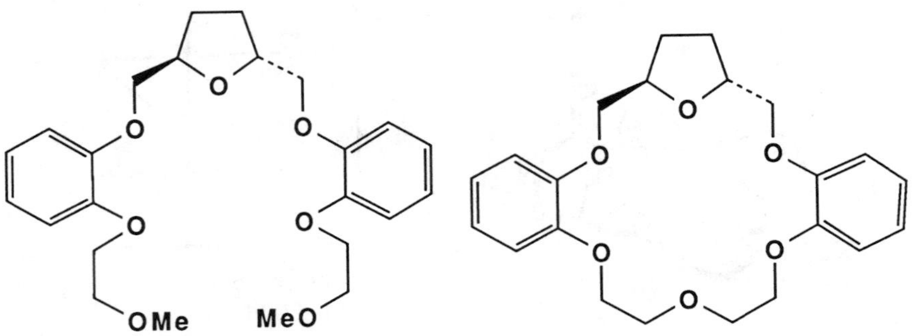

FIGURE 17.   Chiral acyclic and cyclic crown ethers.

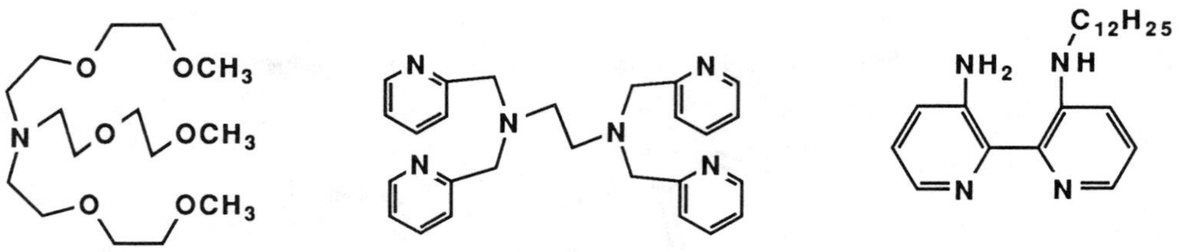

FIGURE 18.   Acyclic crown ethers for transport of heavy metal cations.

and nitrogen atoms in metal coordination. The rate constants of complexation and decomplexation were determined for $K^+$ complex and $1.1 \times 10^8$ $(M^{-1}s^{-1})$ as $4 \times 10^3$ $(s^{-1})$, which are the same orders as those of several naturally occurring ionophore systems.

Tsukube applied this acyclic crown compound to the transport of organic ammonium cations of biogenetic amines.[44] It effectively transported these cations, while $K^+$, $Na^+$, $NH_4^+$, and other inorganic cations were rarely carried. As observed in $Rb^+$ complex, cation-coordination of quinoline nitrogen atoms may induce successive binding of the polyether sequence. Pyridine-bearing polyether carriers offered similar cooperative cation-binding and effective transport of organic ammonium cations.[45] Naemura et al. prepared chiral acyclic crown ether carriers which incorporated chiral trans-2,5-disubstituted tetrahydrofuran subunits (Figure 17).[46] They compared enantiomer-selective transport ability of the acyclic crown compound with that of corresponding cyclic one. Although a direct comparison was difficult, they recorded comparable enantiomer-selectivity for bulky, chiral ammonium cations to that of the crown compound.

Acyclic crown ether carriers sometimes showed excellent transport functions for heavy metal ions as well as organic guests. Tsukube et al. revealed that a tripode with three polyether arms acted as a specific carrier of $Ag^+$ ion in a $CHCl_3$ liquid membrane system (Figure 18).[47] They also examined pyridine-functionalized tetrapode in which a series of nitrogen donors was arranged on a flexible open-chain skeleton. This formed a three-dimensional complex with $Pb^{2+}$ ion and effectively carried it.[48] Kishi et al. reported uphill transport of $Cu^{2+}$ ion by a 6,6'-diamino-2, 2'-bipyridine derivative (Figure 18).[49] This has the ability to transport $Cu^{2+}$ ion against its concentration gradient when a proton gradient is available. Although we know limited examples of excellent acyclic crown ether carriers, this type of compound can offer valuable carrier activities in liquid membrane transport.

## IV. NEW SYNTHETIC CARRIER FOR ANION TRANSPORT

In contrast to a large variety of synthetic carriers for cationic guests, only a limited number of synthetic "anion transport carriers" have been designed. Since amino acids,

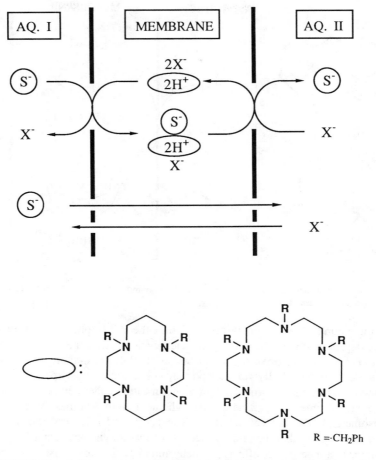

FIGURE 19. Anion transport mediated by polyammonium macrocycle; $S^-$: guest anion, $X^-$: counter-transported anion, $H^+$: proton.

nucleic acids, and other anions are guests of biological and practical interest, a new class of anion-transport carriers should have potential to stimulate exciting developments both in basic and applied liquid membrane sciences. In a similar fashion to the design of cation-transport carriers, synthetic carriers of anionic guests can be designed by considering

1. Size- and shape-fitting of guest and carrier
2. Nature of anion-binding interaction
3. Guest coordination geometry

Macrocyclic polyammonium cations, protonated polyamine macrocycles, can show unique anion-binding properties which are dependent on size and shape of macrocycle. Metal cations may specifically form "ligand-metal ion-guest anion" type ternary complexes. Therefore, new and potentially excellent anion-transport carriers can be derived from these compounds.

## A. POLYAMMONIUM MACROCYCLE

Lipophilic polyammonium macrocycles effectively solubilized anionic guests into the nonpolar media and selectively carried them across a liquid membrane (Figure 19).[50] After being protonated at the Aq.I/membrane interface, the polyamine macrocycle (protonated form) binds the guest anion and transports it through the membrane. At the membrane/Aq.II

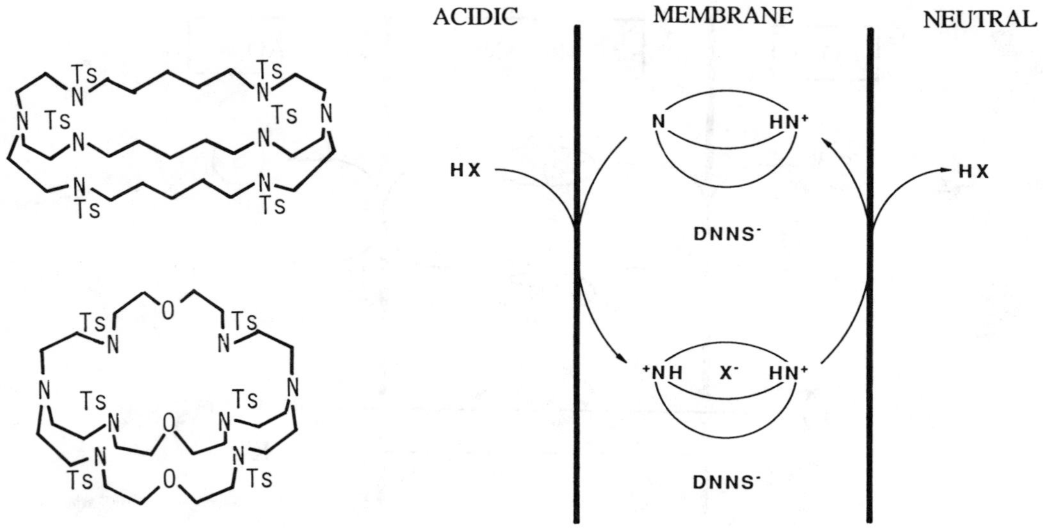

FIGURE 20.   Anion transport mediated by macrobicyclic diammonium salts; X⁻: guest anion, H⁺: proton, DNNS⁻: dinonylnaphthalenesulfonate.

interface, the bound guest anion is released into the Aq.II phase, together with anion-exchange or deprotonation of the polyamine macrocycle. Hence, the antiport-anion or proton concentration gradient can drive the anion transport. Polyammonium macrocycles showed the great advantages expecially for transport of polyanionic guest species. Typically, 14-membered tetra-amine easily formed a diammonium cation under neutral or acidic conditions and transported o-isomer of benzene-dicarboxylic acid 10 to 30 times more effectively than m- and p-isomers. It also favored small benzene-1,2,3- and 1,2,4-tricarboxylic acid anions by 20 to 30 times rather than large benzene-1,3,5-tricarboxylic acid anion. The ring-size of the polyammonium macrocycle effectively determined carrier functions: 18-membered hexa-amine transported larger pyromellitic acid anion more effectively than 14-membered tetra-amine, while hexa-amine offered low transport rates for small di- and tri-carboxylic acid anions. Under the same conditions, quarternary ammonium surfactants such as trioctyl-methylammonium chloride offered "flat" transport selectivity for these polyanionic guest species. Similar dicarboxylate anions are carried by several complicated carrier proteins in bio-membrane transport processes.[51]

Lehn et al. reported that macrobicyclic diammonium salts mediated the artificial symport of Br⁻ anion and proton via "anion cryptate" formation.[52] As shown in Figure 20, the cryptand carrier forms an anion-inclusion complex via protonation/extraction at the acidic interface. At the neutral interface, deprotonation enforces the release of guest species (HX). The large lipophilic anion, dinonylnaphthalene sulfonate (DNNS⁻), can remain outside the carrier cavity, but can stay in the membrane phase. The employed cryptands exhibited characteristic guest selectivities in the solution complexation process which are governed by a combination of electrostatic factors and ligand-anion complementarity of shape and size. Since their transport selectivities were modestly suppressed, the guest anion was sometimes located outside of the cavity as a simple ion-pair complex. Further structural modifications of polycyclic carriers may offer new design possibilities for highly specific carriers for anionic guests.

## B. TRANSITION METAL COMPLEX

Metal-anion coordination interaction can be useful as the driving force of selective carrier-guest anion complexation. Tsukube et al. noticed that naturally occurring metallo-enzymes

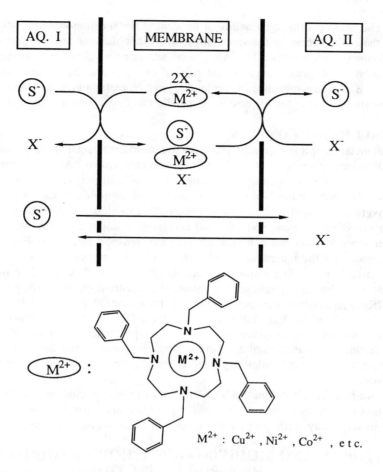

FIGURE 21. Anion transport mediated by transition metal complex carriers; $S^-$: guest anion, $X^-$: counter-transported anion.

specifically recognize and bind the anionic substrates via "ligand protein-central metal cation-guest anion" type complexation. Thus, they prepared several types of lipophilic transition metal complexes as anion-transport carriers.[53-55] For example, transition metal complexes with 1,4,7,10-tetrabenzyl-1,4,7,10-tetraazacyclododecane (tbcyclen) revealed effective carrier functions for anion-transport. Their noteworthy features are

1.  The lipophilic ligand is prepared and modified without difficulty
2.  Various transition metal ions such as copper, nickel, and cobalt ions are available as anion-binding centers
3.  Anion-coordination properties are significantly regulated by choice of polyamine ligand and central metal ions
4.  Substituents of polyamine carrier control hydrophobic/hydrophilic balances of carrier and complex.

Among the examined tbcyclen-metal complexes, its $CuCl_2$ complex most effectively transported several N-benzoyl-amino acid derivatives (carboxylate anions) across a $CHCl_3$ liquid membrane via antiport mechanism (Figure 21). Bz-Ala anion was the most effectively transported guest of all amino acid anions examined: Bz-Ala > Bz-Val > Bz-Gly > Bz-Leu > Bz-Phe. Transport selectivity and efficiency were expectedly modified by changing the central metal ion: $Ni^{2+}$ complex effectively transported Bz-Glu and $Co^{2+}$ complex favored

Bz-Gly. The anion-coordinating ability of the central metal cation was interestingly responsible for these differences in transport properties. When the counter-transported anion was $Cl^-$ ion, concentration ratio of Bz-Ala anions across a tbcyclen-$CuCl_2$ carrier-containing liquid membrane increased from an initial value of 1 to ca. 7 after 24 h under typical transport conditions. Bathophenanthroline, vitamin $B_{12}$, and other lipophilic transition metal complexes are also available for the design of metallo-carriers for anion-transport.

## C. ALKALI METAL COMPLEX

Alkali metal complexes with lipophilic neutral macrocyclic ligands also acted as effective carriers of anionic guest species. Although many neutral crown compounds mediate "symport" of some alkali metal cations, the rates of cation transport are largely influenced by the nature of symport anion (see Figure 6). Lamb et al. proposed that anionic species could be selectively transported with cations by crown ether carriers.[56] Tsukube then investigated typical crown ethers, acyclic podands, and cryptands as anion transport carriers.[57-59]

When dibenzo-18-crown-6 was employed, the transport rate of Bz-Phe (carboxylate) anion depended on the nature and concentration of the cotransported cation. $K^+$ ion was more effective in promoting anion transport than $Na^+$ and $Cs^+$ ions and an increase in its concentration in the Aq. I phase greatly enhanced anion-transport rate. The present symport system offered an interesting transport selectivity for a series of amino acid derivative anions: Bz-Gly < Bz-Ala < Bz-Val < Bz-Leu < Bz-Phe. This is a reversed transport sequence from that observed with transition metal complex-type carriers as metioned above. Other types of crown compounds exhibited similar transport profiles, and their transport abilities increased in the following order: cryptand > crown ether > aza-crown ether > acyclic crown ether.

Now we have two different types of "metallo-carriers" including transition and alkali metal cation centers. Selection of the most appropriate one allows a desired adjustment to be made in selectivity, efficiency, and direction of anion transport process.

# V. NATURAL AND MODIFIED IONOPHORE EXHIBITING NEW TRANSPORT FUNCTION

In carrier chemistry, naturally occurring ionophores are of special interest, though their molecular structures are somewhat complicated. Since their transport functions interestingly resemble those of synthetic carriers as metioned above, we have learned many things about molecular recognition and membrane transport from them. On the other hand, new concepts and methodologies established in synthetic carrier chemistry provide a basis for exploration of a new class of carrier molecules derived from naturally occurring ionophores. Several ambitious approaches are currently developing in the border regions between synthetic and biological carrier chemistry. One of the most exciting approaches of these is structural and/or functional modification of naturally occurring ionophores. This has potential promises not only for understanding of biological membrane transport mechanisms, but also for designing practical applications of natural ionophores. Concentration of valuable or toxic metals, optical resolution of organic compounds, separation of radioisotopes, and related processes may be envisaged using naturally occurring ionophores or their chemically modified derivatives.[60] In this section, we describe new transport functions of naturally occurring ionophores and their chemically modified derivatives, some of which, though not of direct biological significance, are of practical interest.

## A. NATURALLY OCCURRING IONOPHORE

Several naturally occurring ionophores have potential abilities to form stable complexes with natural and unnatural guest cations. Stability constants of some naturally occurring

**TABLE 1**
**Stability Constants of Metal Complexes with**
**Ionophores in Methanol**

| Ionophore | Log K (M$^{-1}$) | | | | |
|---|---|---|---|---|---|
| | Li$^+$ | Na$^+$ | K$^+$ | Rb$^+$ | Cs$^+$ |
| Monensin | — | 4.9 | 4.5 | 4.2 | 3.7 |
| Valinomycin | 0.7 | 1.1 | 4.4 | 6.4 | 5.8 |
| Nonactin | — | 2.7 | 4.5 | 3.8 | 3.2 |
| 18-Crown-6 | 0 | 4.4 | 6.1 | 5.7 | 4.8 |
| Cryptand(2,2,2) | 2.6 | 8.0 | 10.8 | 9.0 | 4.4 |

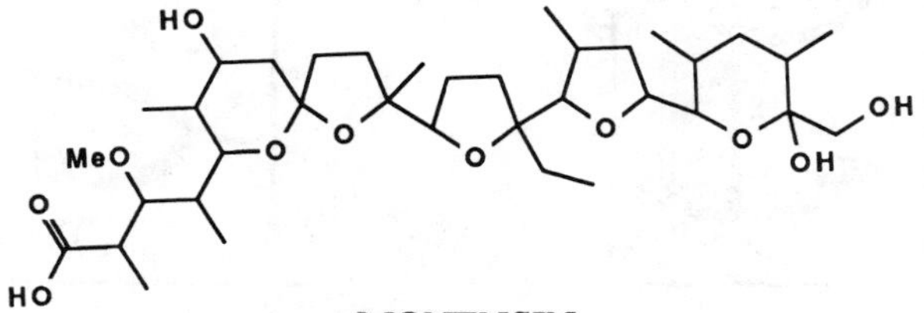

MONENSIN

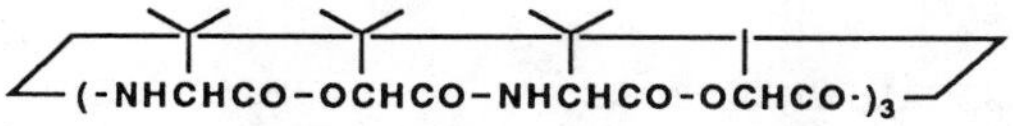

VALINOMYCIN

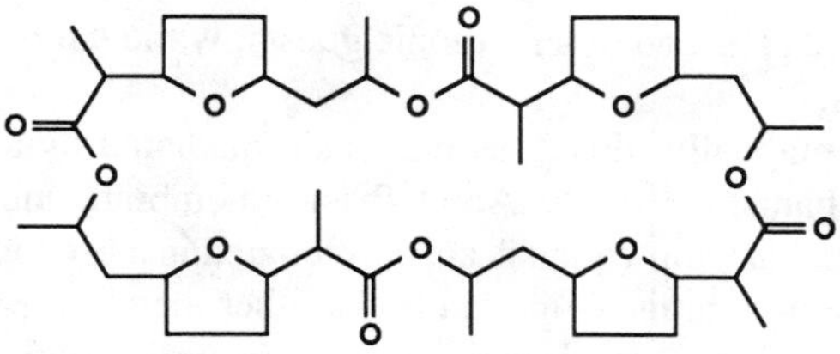

NONACTIN

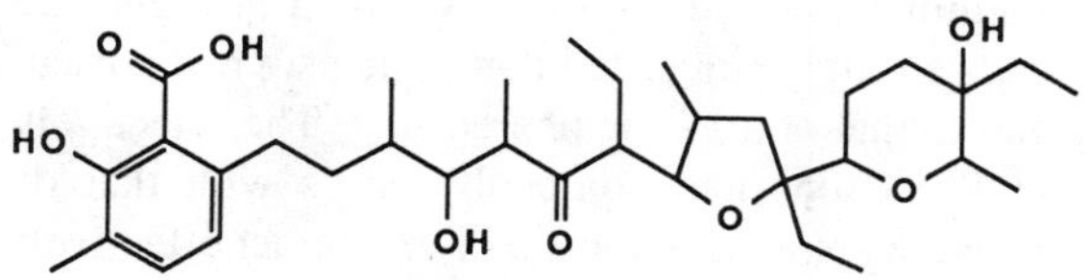

LASALOCID

ionophores are summarized in Table 1, which were measured in methanol solutions.[61] This clearly indicates that these naturally occurring ionophores have ligand arrangements flexible enough to adjust to various guest cations. For instance, monensin typically showed high binding ability for K$^+$, Rb$^+$, and Cs$^+$ ions as well as for biological guest Na$^+$ ion. Thus, it is expected to act as a potential carrier of unnatural guest cations in a synthetic liquid membrane. Tsukube et al. successfully formulated two typical biomimetic transport systems using some biological ionophores.[11] Under nonbiological conditions, monensin mediated

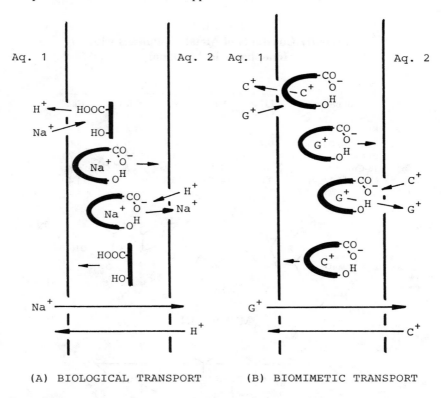

(A) BIOLOGICAL TRANSPORT      (B) BIOMIMETIC TRANSPORT

FIGURE 22. Monensin-mediated biological and biomimetic transport; $G^+$: guest cation, $C^+$: counter-transported cation.

"antiport" of amino acid salts and other cationic guests, while valinomycin mediated "symport" of anionic guests.

Figure 22 (A) schematically illustrates monensin-mediated biological transport of $Na^+$ ion via an antiport mechanism.[62] At the Aq. 1 (basic)/membrane interface, monensin forms an ion-pair complex with $Na^+$ ion which is accommodated in a pseudo-cavity. This lipophilic complex taken up in the membrane is moved to the other membrane side. At the membrane/Aq. 2 (acidic) interface, the $Na^+$ ion is released after neutralization. The cycle is completed by the back transport of the free monensin (neutral form).

On the other hand, monensin effectively transported several amino acid ester salts, $Ag^+$, and $Pb^{2+}$ ions via a biomimetic antiport system as shown in Figure 22 (B).[63] Examination of CPK molecular models strongly suggested that monensin had a pseudo-cyclic cavity, the size of which was similar to that of a 17-membered ring. The effectively transported organic ammonium, $Ag^+$, and $Pb^{2+}$ ions had compatible sizes with that of the pseudo-cavity. Transition metal cations were apparently too small to interact effectively with oxygen atoms of the monensin polyether and were hardly transported. The nature of counter-transported cation had a great influence on overall transport rates, and $Na^+$ ion was the most effective among the alkali metal cations. Since biological lasalocid (see footnote to Table 1) exhibited high transport rates of $Ag^+$, $Ni^{2+}$, $Zn^{2+}$, and $Co^{2+}$ ions, the choice of the employed biological ionophore is an essential factor in altering the profile of this biomimetic transport system.

Macrocyclic valinomycin and nonactin are specific ionophores for biological $K^+$ ion transport. They were also confirmed to mediate biomimetic symport of several amino acid derivative anions and symport metal cations in the same direction (see Figure 6).[64] In particular, valinomycin showed higher transport efficiencies than nonactin and synthetic crown ether carriers. Its transport rate was significantly determined by the natures of guest

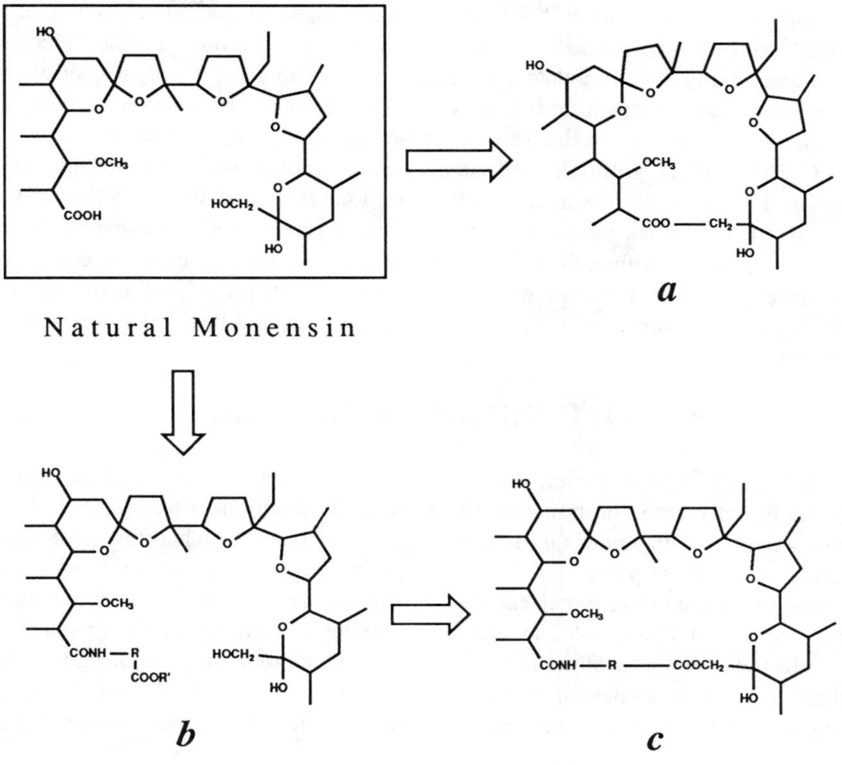

Natural Monensin

*a*

*b*          *c*

FIGURE 23.   Chemical modification of natural monensin.

anion and cotransported cation. Optimal combination of Cs$^+$ and Bz-Phe carboxylate anion provided high transport efficiency. Since other biological and synthetic neutral carriers had favorable combinations of guest salts, the selectivity of this biomimetic amino acid transport system could be flexibly adjusted. In contrast, bilogical valinomycin and nonactin ionophores rarely transported organic ammonium cations of amino acid derivatives, though some synthetic crown compounds offered fast transport. As expected from the X-ray data, the bulky residues of these natural ionophores might prevent tight host-guest complexations with guest ammonium cations of bulky organic guests. Naturally occuring ionophores are not always "infallible" in the synthetic liquid membranes.

## B. CHEMICALLY MODIFIED IONOPHORE

Since naturally occurring ionophores can act as potential carriers for unnatural guest species in the synthetic liquid membranes, they were chemically modified so that further interesting carrier functions might be developed. Although few sucessful examples have yet been reported, the structural modification of naturally occurring ionophore provides a new and promising methodology for use in designing a new class of carrier molecules.[6,60]

Monensin has a particularly interesting structure for chemical modification, which has a pseudo-cavity for guest accommodation encircled by reactive −OH and −COOH groups as well as 17 asymmetric carbons (Figure 23). Corey et al. first prepared a macrocyclic monensin derivative (type a) in which terminal −OH and −CO$_2$H groups were linked via lactone formation.[65] Suzuki et al. recently examined the cation binding profile of this macrocycle using the ion-selective electrode method[66] and found that it showed a good Li$^+$ ion selectivity. Probably, this macrocycle may have a reduced cavity for guest accommodation, and the cation binding selectivity was altered so that it is different from that of natural acyclic monensin.

Naturally occurring monensin ionophore can be used in optical resolution of organic racemates.[67] Since it is composed of 17 asymmetric segments, it may provide a chiral ordered cavity to specifically accommodate certain chiral guest species. Maruyama et al. reported that some modified monensin derivatives (type b) exhibited effective chiral recognition in a liquid membrane-type electrode. The monensin derivatives having neutral end-groups such as $-NH-C*H(CH_3)C_6H_5$ group showed satisfactory enantiomer-selectivity for several racemic amine salts. They favored S-isomers of PheOMe, LeuOMe, and phenylglycine methyl ester salts (almost 4 to 8 times over R-isomers), whereas parent monensin scarcely recognized these optical isomers. Although the examined monensin derivatives show comparable enantiomer-selectivities to those of chiral crown ethers, other kinds of naturally occurring ionophores such as ionomycin, salinomysin, and nigerisin also have other potentials as new chiral host molecules.

## VI. CONCLUDING REMARKS

Synthetic and natural carriers are particularly exciting because they can remarkably modify the liquid membrane transport phenomena. In addition to the traditional chelating reagents, we are finding new families of carrier molecules which offer great promise in broad areas of transport processes. Examples are the facilitated transport of alkali, alkaline earth, heavy and transition metal cations as well as organic guests. Although the carrier syntheses are still challengeable, several useful strategies have led to the design of specific carrier molecules. New concepts and methodologies developing in synthetic carrier chemistry also make it possible to design a new class of carrier molecules derived from naturally occurring ionophores. Carrier chemistry and liquid membrane science are coming of age.

## ACKNOWLEDGMENTS

Finally, I wish to express thanks to Okayama University and Ministry of Education, Science, and Culture, Japan, for providing financial support for our work.

## REFERENCES

1. **Hilgenfeld, R. and Saenger. W.,** Structural chemistry of natural and synthetic ionophores and their complexes with cations, in *Topics in Current Chemistry,* 101, Springer-Verlag, Berlin, 1982, 1.
2. **Painter, G. R. and Pressman, B. C.,** Dynamic aspects of ionophore mediated membrane transport, in *Topics in Current Chemistry,* 101, Springer-Verlag, Berlin, 1982, 83.
3. **Agtarap, A., Chamberlin, J. W., Pinkerton, M., and Steinrauf, L.,** The structure of monensic acid: A new biologically active compound, *J. Am. Chem. Soc.,* 89, 5373, 1967.
4. **Duax, W. L., Smith, G. D., and Strong, P. D.,** Complexation of metal ions by monensin: Crystal and molecular structure of hydrated and anhydrous crystal forms of sodium monensin, *J. Am. Chem. Soc.,* 102, 6725, 1980.
5. **Cox, D. G., Firman, P., and Schneider, H.,** Sodium ion-proton exchange reactions of the carboxylic acid ionophore monensin, *J. Am. Chem. Soc.,* 107, 4297, 1985.
6. **Tsukube, H.,** Characteristics of new crown compounds, in *Crown Ethers and Analogous Compounds,* Hiraoka, M., Ed., Elsevier, Amsterdam, in press, chap. 3.
7. **Reusch, C. F. and Cussler, E. L.,** Selective membrane transport, *AIChE J.,* 19, 736, 1973.
8. **Okahara, M. and Nakatsuji, Y.,** Active transport of ions using synthetic ionophores derived from macrocyclic polyethers and related compounds, in *Topics in Current Chemistry,* 128, Springer-Verlag, Berlin, 1985, 37.
9. **Potvin, P. G. and Lehn, J. M.,** Design of cation and anion receptors, catalysts, carriers, in *Synthesis of Macrocycles,* Izatt, R. M. and Christensen, J. J., Eds., Wiley-Interscience, New York, 1987, 167.
10. **Cram, D. J.,** The design of molecular hosts, guests, and their complexes, *Angew. Chem. Int. Ed. Engl.,* 27, 1009, 1988.

11. **Tsukube, H.,** Biomimetic membrane transport via designed macrocyclic host molecules, *J. Coord. Chem.*, B-16, 101, 1987.

12. **Dishong, D. M., Diamond, C. J., Cinoman, M. I., and Gokel, G. W.,** Syntheses and cation binding properties of carbon-pivot lariat ethers, *J. Am. Chem. Soc.*, 105, 586, 1983.

13. **Koszuk, J. F., Czech, B. P., Walkowiak, W., Babb, D. A., and Bartsch, R. A.,** Lipophilic crown phosphoric acid monoalkyl esters: Synthesis and solvent extraction, *J. Chem. Soc., Chem. Commun.*, 1504, 1984.

14. **Tsukube, H., Takagi, K., Higashiyama, T., Iwachido, T., and Hayama, N.,** New open-chain cryptands with specific ion transport abilities, *J. Inclusion Phenomena*, 2, 103, 1984.

15. **Tsukube, H., Takagi, K., Higashiyama, T., Iwachido, T., and Hayama, N.,** Specific Ag(l) ion-binding and transport properties of tripode type open-chain cryptand, *J. Chem. Soc., Perkin 1*, 1697, 1987.

16. **Tsukube, H.,** Double armed crown ethers with specific cation transport ability, *J. Chem. Soc., Chem. Commun.*, 315, 1984.

17. **Tsukube, H.,** Ion-selective membrane transport via host-guest chemistry, *Ionics*, 121, 15, 1985.

18. **Gokel, G. W. and Korzeniowski, S. H.,** *Macrocyclic Polyether Syntheses*, Springer-Verlag, Berlin, 1982, 1.

19. **Tsukube, H.,** Specific cation-transport ability of new macrocyclic polyamine compounds, *J. Chem. Soc., Chem. Commun.*, 970, 1983.

20. **Ring, K.,** Some aspects of the active transport of amino acids, *Angew. Chem. Int. Ed. Engl.*, 9, 345, 1970.

21. **Lehn, J. M. and Vierling, P.,** The [18]-$N_3O_3$ aza-oxa macrocycle: A selective receptor unit for primary ammonium cations, *Tetrahedron Lett.*, 21, 1323, 1980.

22. **Tsukube, H., Takagi, K., Higashiyama, T., Iwachido, T., and Hayama, N.,** Lipophilic polyamine and polyamide macrocycles for membrane transport of amino acid esters and related cations, *J. Chem. Soc., Perkin 2*, 1541, 1985.

23. **Uiterwijk, J. M. H. M., Staveren, C. J., Reinhoudt, D. N., Hertog, H. J., Kruise, L., and Harkema, S.,** Macrocyclic receptor molecules for guanidinium cation. 2, *J. Org. Chem.*, 51, 1575, 1986.

24. **Izatt, S. R., Hawkins, R. T., Christensen, J. J., and Izatt, R. M.,** Cation transport from multiple alkali cation mixtures using a liquid membrane system containing a series of calixarene carriers, *J. Am. Chem. Soc.*, 107, 63, 1985.

25. **Kimura, E., Dalimunte, C. A., Yamashita, A., and Machida, R.,** A proton-driven copper(II) ion pump with a macrocyclic dioxotetra-amine: A new type of carrier for solvent extraction of copper, *J. Chem. Soc., Chem. Commun.*, 1041, 1985.

26. **Saji, T. and Kinoshita, I.,** Electrochemical ion transport with ferrocene functionalized crown ether, *J. Chem. Soc., Chem. Commun.*, 716, 1986.

27. **Echeverria, L., Delgado, M., Gatto, V. J., Gokel, G. W., and Echegoyen, L.,** Enhanced transport of Li$^+$ through an organic model membrane by an electrochemically reduced anthraquinone podand, *J. Am. Chem. Soc.*, 108, 6825, 1986.

28. **Shinkai, S., Inuzuka, K., Miyazaki, O., and Manabe, O.,** Cyclic-acyclic interconversion coupled with redox between dithiol and disulfide and its application to membrane transport, *J.Am. Chem. Soc.*, 107, 3950, 1985.

29. **Frederick, L. A., Fyles, T. M., Malik-Diemer, V. A., and Whitefield, D.,** A proton-driven potassium ion pump, *J. Chem. Soc., Chem. Commun.*, 1211, 1980.

30. **Shinkai, S., Ogawa, T., Nakaji, T., and Manabe, O.,** Light-driven ion-transport mediated by a photo-responsive bis(crown ether), *J. Chem. Soc., Chem. Commun.*, 375, 1980.

31. **Irie, M. and Kato, M.,** Photoregulated ion capture and release using thioindigo derivatives having ethyl-enedioxy side groups, *J. Am. Chem. Soc.*, 107, 1024, 1985.

32. **Grandour, R. D., Fronczek, F. R., Gatto, V. J., Minganti, C., Schultz, R. A., White, B. D., Arnold, K. A., Mazzocchi, D., Miller, S. R., and Gokel, G. W.,** Solid-state structural chemistry of lariat ethers and BiBle cation complexes, *J. Am. Chem. Soc.*, 108, 4078, 1986.

33. **Tsukube, H., Takagi, K., Higashiyama, T., Iwachido, T., and Hayama, N.,** Cation-binding properties of armed macrocyclic host molecules and their applications to phase transfer reaction and cation membrane transport, *J. Chem. Soc., Perkin 1*, 1033, 1986.

34. **Tsukube, H., Yamashita, K., Iwachido, T., and Zenki, M.,** Pyridino-armed diaza-crown ethers for specific transport of transition metal cations, *Tetrahedron Lett.*, 29, 569, 1988.

35. **Tsukube, H., Adachi, H., and Morosawa, S.,** Ion selectivity control in ester- and amide-armed diaza-crown ethers, *J. Chem. Soc., Perkin 1*, 89, 1989.

36. **Kimura, K., Sakamoto, H., Kitazawa, S., and Shono, T.,** Novel lithium selective ionophores bearing an easily ionizable moiety, *J. Chem. Soc., Chem. Commun.*, 669, 1985.

37. **Nakatsuji, Y., Kobayashi, H., and Okahara, M.,** Active transport of alkali metal cations: A new type of synthetic ionophore derived from a crown ether, *J. Chem. Soc., Chem. Commun.*, 800, 1983.

38. **Shinkai, S., Ishihara, M., Ueda, K., and Manabe, O.,** On-off-switched crown ether-metal ion complexation by photoinduced intramolecular ammonium group "tail-biting", *J. Chem. Soc., Chem. Commun.,* 727, 1984.

39. **Tsukube, H., Adachi, H., and Morosawa, S.,** Amides-armed azamacrocycles as a new series of synthetic carriers for alkali and alkaline earth metal cations, *J. Chem. Soc., Perkin Commun.,* 1537, 1989; **Tsukube, H., Iwachido, T., and Zenki, M.,** Triazamacrocycle having pyridine-pendant arms as a new Na$^+$ ion-selective ionophore, *Tetrahadron Lett.,* 30, 3983, 1989.

40. **Tsukube, H., Takagi, K., Higashiyama, T., Iwachido, T., and Hayama, N.,** Armed thia-crown ether: A specific carrier for soft metal cation, *Tetrahedron Lett.,* 26, 881, 1985.

41. **Chang, S. K. and Cho, I.,** New metal cation-selective ionophores derived from calixarenes: Their syntheses and ion-binding properties, *J. Chem. Soc., Perkin 1,* 211, 1986.

42. **Kimura, K., Matsuo, M., and Shono, T.,** Lipophilic calix[4]arene ester and amide derivatives as neutral carriers for sodium ion selective electrodes, *Chem. Lett.,* 615, 1988.

43. **Vogtle, F. and Weber, E.,** Multidentate acyclic neutral ligands and their complexations, *Angew. Chem. Int. Ed. Engl.,* 18, 753, 1979.

44. **Tsukube, H.,** Highly selective transport of biogenetic amines and drugs by using functionalized acyclic crown ether, *Tetrahedron Lett.,* 23, 2109, 1982.

45. **Tsukube, H.,** Pyridine-containing acyclic crown ethers and their polymers as specific carriers of biogenetic amine derivatives, *Bull. Chem. Soc. Jpn.,* 55, 3882, 1982.

46. **Naemura, K., Ebashi, I., Matsuda, A., and Chikamatsu, H.,** Enantiomer recognition of crown ethers and open-chain polyethers containing the trans-tetrahydrofuran-2,5-diylbis(methylene)subunit as the chiral centre, *J. Chem. Soc., Chem. Commun.,* 666, 1986.

47. **Tsukube, H., Takagi, K., Higashiyama, T., Iwachido, T., and Hayama, N.,** Tripode-type ionophore for specific transport of Ag(l) ion, *Chem. Lett.,* 1079, 1986.

48. **Tsukube, H., Yamashita, K., Iwachido, T., and Zenki, M.,** Specific Pb$^{2+}$ ion-binding and transport properties of pyridine-armed tetrapode ionophores, *J. Chem. Res. (S),* 104, 1988.

49. **Kishi, N., Araki, K., and Shiraishi, S.,** The specific and uphill transport of copper(II) ion by a 6,6'-diamino-2,2'-bipyridine derivative, *J. Chem. Soc., Chem. Commun.,* 103, 1984.

50. **Tsukube, H.,** Lipophilic macrocyclic tetraamine as specific carrier of amino acid and related anions, *Tetrahedron Lett.,* 24, 1519, 1983. **Tsukube, H.,** Proton-assisted transport of amino acid and related polycarboxylate anions via polyammonium macrocycles, *J. Chem. Soc., Perkin 1,* 615, 1985.

51. **Lo, T. C. Y.,** The transfer of a bacterial transmembrane function to eukaryotic cells, *J. Biol. Chem.,* 254, 591, 1979.

52. **Dietrich, B., Fyles, T. M., Hosseini, M. W., Lehn, J. M., and Kaye, K. C.,** Proton coupled membrane transport of anions mediated by cryptate carriers, *J.Chem. Soc., Chem. Commun.,* 691, 1988.

53. **Tsukube, H.,** Active transport of amino acid anions by a synthetic metal complex carrier, *Angew. Chem. Int. Ed. Engl.,* 21, 304, 1982.

54. **Maruyama, K., Tsukube, H., and Araki, T.,** Carrier-mediated transport of amino acid and simple organic anions by lipophilic metal complexes, *J. Am. Chem. Soc.,* 104, 5197, 1982.

55. **Tsukube, H.,** Artificial transport of amino acid, oligopeptide, and related anions by macrocyclic polyamine-transition metal complex carriers, *J. Chem. Soc., Perkin 1,* 29, 1983.

56. **Lamb, J. D., Christensen, J. J., Izatt, S. R., Bedke, K., Astin, M. S., and Izatt, R. M.,** Effects of salt concentration and anion on the rate of carrier-facilitated transport of metal cations through bulk liquid membranes containing crown ethers, *J. Am. Chem. Soc.,* 102, 3399, 1980.

57. **Tsukube, H.,** K$^+$ ion dependent, active transport of amino acid anions by macrocyclic carriers, *Tetrahedron Lett.,* 22, 3981, 1981.

58. **Tsukube, H.,** Active and passive transport of amino acid and oligopeptide derivatives by artificial ionophore-K$^+$ complexes, *J. Chem. Soc., Perkin 1,* 2359, 1982.

59. **Tsukube, H.,** Noncyclic crown-type polyether polymers for transport of alkali cation and amino acid anion, *J. Polym. Sci., Polym. Chem. Ed.,* 20, 2989, 1982.

60. **Tsukube, H.,** Cation-binding by natural and modified ionophores: From natural ionophore to man-made ionophore, in *Cation-Binding by Macrocycles,* Inoue, Y. and Gokel, G. W., Eds., Marcel Dekker, New York, in press.

61. **Cox, B. G., Trung, N., Rzeszotarska, J., and Schneider, H.,** Rates and equilibria of alkali metal and silver ion complex formation with monensin in ethanol, *J. Am. Chem. Soc.,* 106, 5965, 1984.

62. **Choy, E. M., Evans, D. F., and Cussler, E. L.,** A selective membrane for transporting sodium ion against its concentration gradient, *J. Am. Chem. Soc.,* 96, 7085, 1974.

63. **Tsukube, H., Takagi, K., Higashiyama, T., Iwachido, T., and Hayama, N.,** Biomimetic transport of unusual metal cations and amino acid ester salts mediated by monensin and lasalocid A, *J. Chem. Soc., Chem. Commun.,* 448, 1986.

64. **Tsukube, H., Takagi, K., Higashiyama, T., Iwachido, T., and Hayama, N.,** Biomimetic application of natural valinomycin and nonactin ionophores to artificial transport of amino acid salts, *Bull. Chem. Soc. Jpn.,* 59, 2021, 1986.

65. **Corey, E. J., Nicolau, K. C., and Melvin, L. S.,** Synthesis of novel macrocyclic lactones in the prostaglandin and polyether antibiotic series, *J. Am. Chem. Soc.,* 97, 653, 1975.

66. **Suzuki, K., Tohda, K., Sasakura, H., Inoue, H., Tatsuta, K., and Shirai, T.,** Natural carboxylic polyether derivatives as lithium ionophores, *J. Chem. Soc., Chem. Commun.,* 932, 1987.

67. **Maruyama, K., Sohimiya, H., and Tsukube, H.,** New chiral host molecules derived from naturally occurring monensin ionophore, *J. Chem. Soc., Chem. Commun.,* 864, 1989.

Chapter 5

# MECHANISTIC PRINCIPLES OF LIQUID MEMBRANE TRANSPORT

**Yoshihisa Inoue**

## TABLE OF CONTENTS

# I. INTRODUCTION

Selective transports of specific neutral and charged species across bio- and artificial membranes are not only essential for the cell metabolism to maintain life, but they also enjoy wide practical applications to the separation science. However the mechanistic and energetic details of the transport phenomena do not appear to be fully clarified. Several transport models have been proposed and treated theoretically in order to elucidate the passive and active transport systems.[1-7] The theories derived are useful in explaining the existing systems and also in designing more selective and efficient transport systems.

Membranes may be classified into two categories: (1) polymer membranes, which exhibit relatively low transmembrane fluxes and poor substrate selectivities, but possess high structural strength and durability, and (2) liquid membranes which in general show higher fluxes and selectivities/specificities and are suitable for mechanistic investigations. Liquid membranes, which we will discuss in this chapter, may be prepared in apparently different forms as follows: (1) bulk liquid; (2) supported liquid film; (3) emulsion; and (4) vesicle. It is important and inevitable to establish the fundamental kinetics which are applicable commonly to a wide variety of membrane configurations. We therefore start with simple diffusion of neutral and charged species in the condensed media and discuss the fundamental equations applicable to the later discussion on the transmembrane phenomena. Then, the two major categories of the membrane transport, passive and active transport, are discussed from the mechanistic point of view.

# II. DIFFUSION

## A. NEUTRAL SPECIES — FICK'S FIRST LAW

In formal analogy with the Ohm's law for electric current or the Fourier's law of heat flow, where the applied potential difference or the temperature gradient is the driving force, respectively, the diffusion of particles in the condensed media is driven by the difference in chemical potential. Under a concentration or, more strictly, activity gradient $dC_i/dx$ along the x axis, the flux of a neutral species i in that direction, i.e., $J_i$, is known to obey the Fick's first law:

$$J_i = -D_i dC_i/dx \tag{1}$$

where the proportionality constant $D_i$ refers to the diffusion coefficient. This equation well describes the diffusion processes of any neutral species and also of charged species as far as no applied potential is present.

## B. CHARGED SPECIES
### 1. Nernst-Planck Equation

When the concentration gradient is the only driving force to produce flux, the motion of charged species obeys the Fick's first law as above. However, a new force-flux equation is to be derived for the motion of charged species under a potential applied. The gradient of electrochemical potential, in place of concentration, becomes an effective driving force. The electrochemical potential is a function of electric potential $\phi$ and concentration (activity) $C_i$ as well as the temperature and pressure of the system, both of which are constant in most cases:

$$\mu_i = \mu_i^\circ(T,P) + z_i F\phi + RT\ln C_i \tag{2}$$

where $z_i$ and F refer to the valency of ion and the Faraday constant, respectively.

The force associated with the electrochemical potential, working upon an ion i, is given by $X_i$:

$$X_i = -(\partial\mu_i/\partial x)_{T,P}$$

$$= -(z_iF(d\phi/dx) + RT(dlnC_i/dx))$$

$$= -z_iF(d\phi/dx) - (RT/C_i)(dC_i/dx) \tag{3}$$

The flux of ion is proportional to the force and is expressed by the Nernst-Planck equation:

$$J_i = u_iC_iX_i$$

$$= -z_iFu_iC_i(d\phi/dx) - u_iRT(dC_i/dx) \tag{4}$$

where $u_i$ refers to the mobility of ion.

In the absence of electrochemical potential, Equation 4 is identical to the Fick's first law. Then we obtain the Einstein Equation (Equation 5) for the diffusion coefficient:

$$D_i = u_iRT \tag{5}$$

The first term of Equation 4 is synonymous with Ohm's law, if the proportionality factor $z_iFu_iC_i$ is taken as conductivity. Thus the Nernst-Planck Equation may be regarded as a linear combination of Ohm's and Fick's laws

The fundamental diffusion equation (Equation 4) for the flux of ions contains two derivative terms concerning the electrical potential and the concentration. This means that we have to know both quantities as a function of position. For a simple solution containing only a uni-univalent electrolyte, both quantities may be calculated. The concentration and potential differences over the entire distance that ion travels are plotted against the distance x ($0 \leq x \leq 1$) in Figure 1, in which the concentration changes linearly in sharp contrast to the curvature for the potential difference. However, it is not always possible to obtain such straightforward results for more complex solutions. Some hypotheses have therefore been developed, assuming either a constant electric field or a constant concentration gradient.

## 2. Goldman Equation for Constant Field

The first hypothesis is the constant field approximation; the electric field is assumed to be held constant over the diffusing distance 1. In other words, the electric field would not be affected by the presence of ions. Being apparently contradictory with the above-mentioned uni-univalent electrolyte case, this approximation gives satisfactory results when the concentration difference is not very large.

Assuming a constant field of $\Delta\phi/l$ over the entire diffusing distance l, the Nernst-Planck Equation is simplified:

$$J_i = -z_iFu_iC_i(\Delta\phi/l) - u_iRT(dC_i/dx) \tag{6}$$

At the steady state, the flux $J_i$ is constant over the diffusing distance and is independent of position x. The differential equation (Equation 6) for a single ion is rearranged with respect to x and $C_i$ and then integrated over the ranges $0 \leq x \leq 1$ and $C_i^0 \leq C_i \leq C_i^1$, respectively. The integration and the subsequent modifications give the Goldman equation for the flux of ion:

$$J_i = -\left(\frac{z_iFu_i\Delta\phi}{1}\right)\frac{C_i^0 - C_i^1\exp(z_iF\Delta\phi/RT)}{1 - \exp(z_iF\Delta\phi/RT)} \tag{7}$$

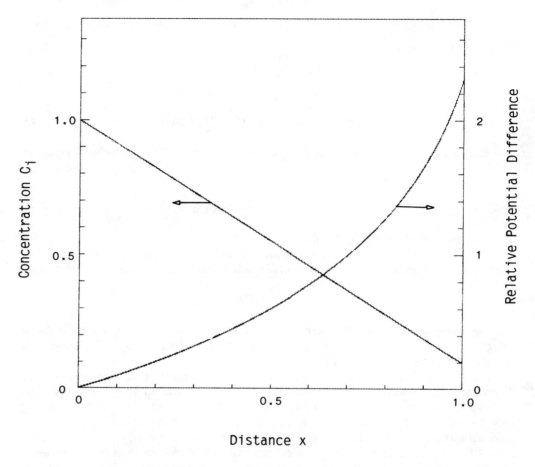

FIGURE 1.    Plots of concentration and potential difference as a function of distance for uni-univalent electrolyte solution.

Some limiting cases may clarify this somewhat complicated equation. With a large positive potential $\Delta\phi$ applied, the exponential terms dominate in both numerator and denominator and the equation reduces to

$$J_i = -z_iFu_iC_i^1\Delta\phi/l \qquad (8)$$

Thus the electrodiffusion of a cation is governed mostly by the concentration at distance 1. The reversed situation is attained when a large negative potential is applied; Equation 7 reduces to

$$J_i = -z_iFu_iC_i^0\Delta\phi/l \qquad (9)$$

It is interesting to consider a situation where the electrodiffusion is approaching the end point, producing a uniform solution. When the potential is small enough, the exponential term may be expressed in a linear form: e.g., $\exp(A) = 1 + A$. Thus the flux equation is described as

$$J_i = (u_iRT/l)(C_i^0 - C_i^1 - z_iFC_i^1\Delta\phi/RT) \qquad (10)$$

At zero potential, the equation reduces to

$$J_i(\Delta\phi \to 0) = -(u_iRT/l)(C_i^1 - C_i^0)$$

$$= -D_i(C_i^1 - C_i^0)/l \tag{11}$$

and is consistent naturally with Equation 1 for the simple diffusion without electrical driving force.

## 3. Constant Concentration Gradient

Another approach to the general solution of the flux Equation 4 is to assume that the concentration gradient $dC_i/dx$ is constant over the entire diffusing region. The constant concentration gradient over the entire diffusing region is realized *a priori* only for the univalent electrolyte solutions as shown in Figure 1, but this is not always the case with a variety of multivalent electrolyte solutions.

The constant concentration gradient throughout the diffusing distance l means

$$dC_i/dx = \Delta C_i/l \tag{12}$$

Integration of Equation 12 gives concentration as a function of distance x.

$$C_i(x) = C_i^0 + \Delta C_i x/l \tag{12a}$$

Thus, the Nernst-Planck Equation is expressed:

$$J_i = -z_iFu_i(C_i^0 + \Delta C_i x/l)(d\phi/dx) - u_iRT\Delta C_i/l \tag{13}$$

Again at the steady state, the flux $J_i$ is constant over the diffusing distance and is independent of x. The differential equation (Equation (13) for a single ion is rearranged with respect to x and $\phi_i$, and then integrated over the ranges $0 \leq x \leq l$ and $\phi_i^0 \leq \phi_i \leq \phi_i^1$, respectively. The integration and the subsequent modifications give the equation:

$$J_i = -(z_iFu_i\Delta\phi\Delta C_i/l)/\ln(C_i^1/C_i^0) - u_iRT\Delta C_i/l \tag{13a}$$

At the limit of zero potential, the equation reduces to Fick's first law.

$$J_i = -u_iRT\Delta C_i/l \tag{14}$$

# III. PASSIVE TRANSPORT

In the passive transport, the transmembrane mass transport is driven by the difference in the chemical potential of the species of interest between two phases separated by a membrane. The mechanism may involve either simple diffusion or facilitated diffusion mediated by a carrier. In contrast to the active transport described later, the flux follows the chemical potential gradient of the species transported.

## A. SIMPLE TRANSPORT

Membranes containing no specific carrier may show simple transport which is governed by the partition of the species to be transported between a membrane and an aqueous phase and also by the diffusion through the membrane. The concentrations $C_i^0$ and $C_i^1$ of the species to be transported just inside the membrane boundaries are determined by the bulk concentrations of the source phase I ($[i]^I$) and the receiving phase II ($[i]^{II}$) separated by the membrane of a thickness of l and by the distribution coefficient $S_i$ between membrane and aqueous phases.

$$S_i = C_i/[i] \quad \text{or} \quad C_i = S_i[i] \tag{15}$$

With neutral species, the transmembrane diffusion obeys Fick's first law as long as the transport is a diffusion-limited process. Thus the flux from the source to the receiving phase through a liquid membrane is expressed by the bulk concentrations [i] and the distribution coefficient $S_i$.

$$J_i = D_i(S_i[i]^I - S_i[i]^{II})/l \tag{16}$$

## B. GOLDMAN-HODGKIN-KATZ EQUATION

The Goldman Equation describes the diffusion of a single ion in the membrane phase as described above. However, under the actual conditions for membrane transport, additional charged species bearing different charge and/or mobility are incorporated in the diffusion process. The Goldman-Hodgkin-Katz Equation was derived from the constant field Goldman Equation in order to describe the relative contributions of $Na^+$, $K^+$, and $Cl^+$ ions to the total charge flux across biomembrane. It is assumed that all ionic species permeate the membrane independently without any interfering or attracting effect from the other ionic species in the membrane.

The total charge flux, i.e., current I, is expressed as a sum of the respective fluxes:

$$I = z_{Na}FJ_{Na} + z_KFJ_K + z_{Cl}FJ_{Cl} \tag{17}$$

Each of the fluxes $J_i$ is given by the Goldman Equation, and Equation 17 is written as

$$I = (F^2\Delta\phi/l)[(u_{Na}C_{Na}^0 + u_KC_K^0 + u_{Cl}C_{Cl}^l) - (u_{Na}C_{Na}^l$$
$$+ u_KC_K^l + u_{Cl}C_{Cl}^0)\exp(F\Delta\phi/RT)]/[1 - \exp(F\Delta\phi/RT)] \tag{18}$$

Since the total current I is equal to zero at the equilibrium,

$$u_{Na}C_{Na}^0 + u_KC_K^0 + u_{Cl}C_{Cl}^l = (u_{Na}C_{Na}^l + u_KC_K^l + u_{Cl}C_{Cl}^0)\exp(F\Delta\phi/RT)$$

then the electrochemical potential $\Delta\phi$ is calculated by the following equation:

$$\Delta\phi = (RT/F)\ln[(u_{Na}C_{Na}^0 + u_KC_K^0 + u_{Cl}C_{Cl}^l)/(u_{Na}C_{Na}^l) + u_KC_K^l + u_{Cl}C_{Cl}^0)] \tag{19}$$

For the membrane transport, the potential is expressed in terms of the bulk concentrations [i] of ions in the aqueous phase by using the distribution coefficient $S_i$ between the membrane and aqueous phase ($S_i = C_i/[i]$):

$$\Delta\phi = (RT/F)\ln\left(\frac{u_{Na}S_{Na}[Na]^I + u_KS_K[K]^I + u_{Cl}S_{Cl}[Cl]^{II}}{u_{Na}S_{Na}[Na]^{II} + u_KS_K[K]^{II} + u_{Cl}S_{Cl}[Cl]^I}\right)$$
$$= (RT/F)\ln\left(\frac{P_{Na}[Na]^I + P_K[K]^I + P_{Cl}[Cl]^{II}}{P_{Na}[Na]^{II} + P_K[K]^{II} + P_{Cl}[Cl]^I}\right) \tag{20}$$

where $P_i$ is the permeability of each ionic species and is equal to $u_iRT/l$.

Although the above Goldman-Hodgkin-Katz Equation was originally developed for biomembrane, it may be applied to a wide variety of membrane systems.

( a )

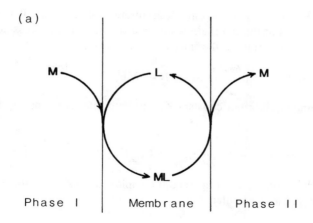

Phase I     Membrane     Phase II

( b )

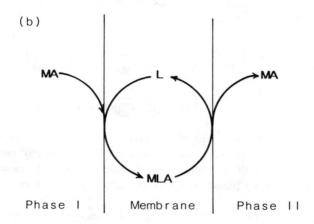

Phase I     Membrane     Phase II

FIGURE 2. (a) Facilitated transport of M with a carrier L through neutral complex formation ML. (b) Facilitated transport of an ion pair MA with a carrier L through neutral complex formation MLA.

## C. FACILITATED TRANSPORT

The transport of specific species through biomembranes has been shown to be mediated by an ionophore to produce highly selective flow of the species, which has stimulated the investigations of the facilitated transport through artificial membrane systems.

Usually charged metal ions are impermeable in a nonpolar liquid membrane, and the concentration gradient between the two aqueous phase separated by a membrane produces no appreciable flow. In the presence of a synthetic ligand or natural ionophore which forms a lipophilic cation-ligand complex, the transport of cation and accompanying anion, if any, is highly facilitated. Moreover the selective transport of ionic species may be materialized in principle by using the ion selective ligands like crown ethers or antibiotic ionophores.

In the facilitated transport, the flux is affected by the substrate concentration, the complexation constant between substrate and carrier, the carrier concentration, the diffusion coefficient, and the thickness of the membrane.

We first consider a simple example in which a carrier L and a species M form a neutral complex at the membrane interface, and the complex produced diffuses across the membrane as shown in Figure 2(a), in which the bulk concentration of M in each aqueous phase is

$C_M^I$ and $C_M^{II}$ ($C_M^I > C_M^{II}$) and the total concentration of L is $C_L^t$. If the complexation/decomplexation process at the interface is fast and the diffusion process across membrane is rate-limiting, the equilibrium at the interface is expressed as

$$M + L \rightleftarrows ML \tag{21}$$

Introducing the same equilibrium constant $K_{ML}$ at both interfaces and the free ligand concentration $C_L$,

$$K_{ML} = C_{ML}/C_M C_L = C_{ML}/C_M(C_L^t - C_{ML}) \tag{22}$$

Then the concentration of the resulting neutral complex just inside the membrane contacting with each aqueous phase is calculated as

$$C_{ML} = K_{ML}C_M C_L^t/(1 + K_{ML}C_M) \tag{23}$$

Using Fick's first law (Equation 1), the flux is expressed by a modification of the Michaelis-Menten-type equation for enzymatic reactions:

$$
\begin{aligned}
J_M &= \frac{D_{ML}K_{ML}C_L^t}{1} \left( \frac{C_M^I}{1 + K_{ML}C_M^I} - \frac{C_M^{II}}{1 + K_{ML}C_M^{II}} \right) \\
&= \frac{D_{ML}K_{ML}C_L^t}{1} \left( \frac{C_M^I - C_M^{II}}{(1 + K_{ML}C_M^I)(1 + K_{ML}C_M^{II})} \right)
\end{aligned}
\tag{24}
$$

where $D_{ML}$ is the diffusion coefficient of ML and l is the effective diffusion path length across the membrane. The equation indicates that the flux is directly proportional to the total carrier concentration in the membrane and that the transport ceases when the concentrations become same in both phases as far as the same equilibrium constant is applied at both interfaces as assumed.

At low $C_M$ and/or small $K_{ML}$, where $K_{ML}C_M \ll 1$, Equation 24 reduces to

$$J_M = D_{ML}K_{ML}C_L^t(C_M^I - C_M^{II})/l \tag{25}$$

while at high $C_M$, where $K_{ML}C_M \gg 1$, the flux approaches its maximum value $J_{max}$

$$J_{max} = D_{ML}C_L^t/l \tag{26}$$

which is independent of the substrate concentrations or the equilibrium constant $K_{ML}$. Thus the initial transport rate $J_{ML}$ is proportional to the concentration difference between the two aqueous phases at low concentrations, but gives a curvature approaching $J_{max}$ in sharp contrast to the simple transport; see Figure 3 (upper trace).

The transport of an ion pair MA mediated by a neutral carrier L leads to different kinetics owing to the concurrent transport of accompanying anion; see Figure 2(b). The equilibrium at the interface may be described by the successive reactions, i.e., the ion pair partitioning of an aqueous metal salt MA to the membrane and the subsequent complexation with a carrier L forming a neutral ion pair complex MLA in the membrane.*

---

\* However, an a.c. polarographic study has suggested that in the nitrobenzene-water system the complexation of sodium ion with a crown ether does not take place in the aqueous or organic phase but at the interface (Reference 8).

85

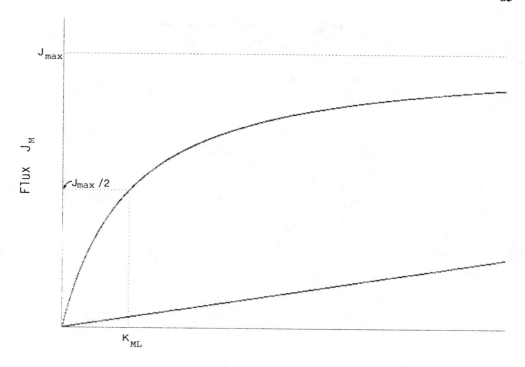

FIGURE 3. Schematic drawing of the initial flux as a function of concentration difference between the source and receiving phases in simple (lower trace) and facilitated transport (upper trace).

$$M^+ + A^- \xrightleftharpoons{K_{MA}} MA \tag{27}$$

$$MA + L \xrightleftharpoons{K_{MLA}} MLA \tag{28}$$

The flux for cotransport of an ion pair is expressed as follows.

$$
\begin{aligned}
J_M &= \frac{D_{MLA}K_{MA}K_{MLA}C_L^t}{l} \left( \frac{(C_M^I)^2}{1 + K_{MA}K_{MLA}(C_M^I)^2} - \frac{(C_M^{II})^2}{1 + K_{MA}K_{MLA}(C_M^{II})^2} \right) \\
&= \frac{D_{MLA}K_{ex}C_L^t}{l} \left( \frac{(C_M^I)^2 - (C_M^{II})^2}{(1 + K_{ex}(C_M^I)^2)(1 + K_{ex}(C_M^{II})^2)} \right)
\end{aligned} \tag{29}
$$

where $D_{MLA}$ is the diffusion coefficient of the ion pair complex MLA and $K_{ex}$ is the overall extraction equilibrium constant: $K_{ex} = K_{MA}K_{MLA}$.

When the bulk concentration of $M^+$ in the source phase I is low and that in the receiving phase II is negligible, Equation 29 reduces to the flux equation of Reusch and Cussler[9]:

$$
\begin{aligned}
J_m &= D_{MLA}K_{MA}K_{MLA}C_L^t(C_M^I)^2/l \\
&= D_{MLA}K_{ex}C_L^t(C_M^I)^2/l
\end{aligned} \tag{30}
$$

The equation derived indicates that the flux is proportional to equilibrium constant $K_{MLA}$,

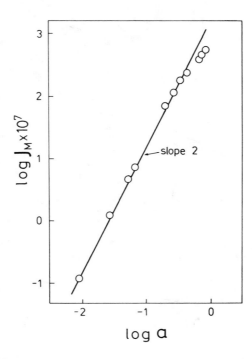

FIGURE 4.    Cation flux as a function of activity for the transport of potassium nitrate across chloroform membrane containing dibenzo-18-crown-6 ($7 \times 10^{-4} \ M$). (Adapted from reference 12.)

ligand concentration $C_L^t$, and square of bulk cation concentration $C_M$. This equation has been examined with respect to cation concentration (activity), ligand concentration, and anion type to give satisfactory results at low cation activities,[9-14] as exemplified in Figure 4.[12]

Although the Reusch-Cussler model also predicts correctly the linear relationship with respect to the equilibrium constant only at low $K_{MLA}$, the model fails to predict the appearance of a peak flux at specific $K_{MLA}$ and the subsequent decrease at higher $K_{MLA}$[11,15]; see Figure 5. Izatt and co-workers have developed a new model for the diffusion-limited membrane transport which involves double diffusion layers just inside the membrane phase as shown in Figure 6(c).[11,16] Although simpler models (Figure 6(a) and (b), which assume that the relevant reaction equilibria occur at the interface, cannot account for the sharp peak in Figure 5, their somewhat sophisticated double-layer model well simulates the above-mentioned behavior exhibiting a sharp peak in the flux at a specific $K_{MLA}$ value. The experimental data well fit the curves in Figure 5, which are predicted by using a much more complicated equation.[11,16]

However, more general and comprehensive theoretical treatments of the diffusion-limited membrane transport mediated by a neutral or charged carrier has been developed by Behr et al.[17] They performed computer-simulations of the flux equations (Equation 24) and (Equation 29) for the facilitated transmembrane diffusion of a single species or an ion pair with a neutral carrier and also of the flux equation for the facilitated countertransport of two charged species with a charged carrier. Their models are quite simple and postulate only one diffusion layer at the interface (Figure 6(a)) rather than the multidiffusion layers (Figure 6(c)), but still give satisfactory results in reproducing the peak flux at specific log $K_{ex}$ under the conditions frequently employed. In view of their simulation, several reported data on the membrane transport[18-31] have been examined to confirm that the transport rate is not limited by the complexation kinetics but by the diffusion process.[17] They further showed

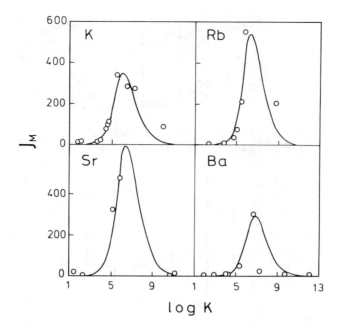

FIGURE 5.   Plot of the flux $J_M$ as a function of log $K_{MLA}$ ($CH_3OH$) in the transport of some alkali and alkaline earth metal salt across chloroform membrane mediated by dibenzo-18-crown-6. (Adapted from reference 11.)

unequivocally that the real transport selectivities between two or more species are evaluated not from the relative rates determined in separate experiments, but by true competition experiment. This has been demonstrated experimentally by Izatt et al. in the membrane transport of binary and multiple cation mixtures.[29,30,32-40]

Detailed experimental studies on the origin of the peak flux have been carried out by measuring the apparent rates of uptake and release independently in the membrane transport of metal picrates with a variety of neutral carriers.[41-44] The observed peak in the flux has been rationalized as a consequence of the combined effect of the slow-uptake/fast-release in the low $K_{ex}$ region and the fast-uptake/slow-release in the high $K_{ex}$ region.

More recently, the transports of potassium perchlorate through supported liquid membranes have been investigated, and the diffusion through the membrane has been shown to be a rate-determining step.[45]

## IV. ACTIVE TRANSPORT

Although the transport selectivity of biomembrane would be mimicked in part by the passive transport facilitated by some synthetic carriers, the actual transport through biomembrane is frequently coupled with the flow of other substance or the chemical reaction. In this section, we will describe the mechanistic principles of the active transport.

### A. IRREVERSIBLE THERMODYNAMICS

The active transport phenomena involve the concurrent coupled flow of several species in the system and are analyzed by irreversible thermodynamics. When the flux $J_i$ of a species i is driven by the conjugate force $X_i$, we can define the dissipation function ($\Phi$) for each species as a product of them. The dissipation function is equal to the product of the temperature T and the rate of entropy generation dS/dt in the system and is therefore greater than zero:

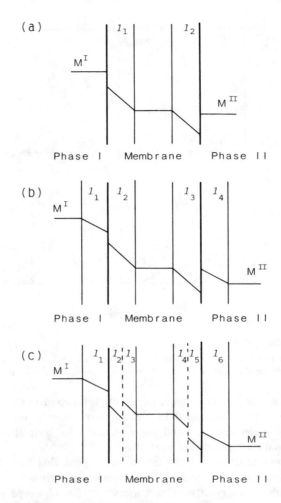

FIGURE 6.   The detailed descriptions of the diffusion layer and the concentration gradients in three models of the diffusion-limiting membrane transport.

$$\Phi = T(dS/dt) = \Sigma J_i X_i > 0 \tag{31}$$

In a typical active membrane transport system in which transport of species i and j across membrane and a chemical reaction r take place concurrently, the dissipation function is written as

$$\Phi = J_i X_i + J_j X_j + J_r X_r \tag{32}$$

The flux $J_i$ is produced not only by its conjugate force $X_i$ but also by the coupled forces $X_j$ and $X_r$. For the three conjugated forces and fluxes, the linear equations can be written as

$$J_i = L_{ii} X_i + L_{ij} X_j + L_{ir} X_r \tag{33}$$

$$J_j = L_{ji} X_i + L_{jj} X_j + L_{jr} X_r \tag{34}$$

$$J_r = L_{ri} X_i + L_{rj} X_j + L_{rr} X_r \tag{35}$$

where Ls are called conductance coefficients which represent the degrees of mutual coupling between the forces indicated. For these equations, the Onsager reciprocal relation holds:

$$L_{ij} = L_{ji} \tag{36}$$

$$L_{ir} = L_{ri} \tag{37}$$

$$L_{jr} = L_{rj} \tag{38}$$

Since $\Phi > 0$, all Ls are positive and $L_{ii}L_{jj} \geq L_{ij}^2$, $L_{ii}L_{rr} \geq L_{ir}^2$, and so on.

In analogy with Equation 33, the force $X_i$ is expressed as a linear combination of the fluxes:

$$X_i = R_{ii}J_i + R_{ij}J_j + R_{ir}J_r \tag{39}$$

where Rs are resistance coefficients. The equation is rewritten as

$$J_i = (1/R_{ii})X_i - (R_{ij}/R_{ii})J_j - (R_{ir}/R_{ii})J_r \tag{40}$$

The first term of the equation represents the contribution of the conjugate force $X_i$ to the flux of species i, the second represents that of the coupled flux of species j, and the third represents that of the chemical reaction r.

Although the term "active transport" was originally defined by Rosenberg[46] as a phenomenon in which the flux opposite to its conjugate driving force is observed, Kedem[47] has proposed a more general definition on the basis of the Equation 40. The passive transport is related to the first term, while the second and third terms are associated with the active transport and are called osmo-osmotic coupling and chemi-osmotic coupling, respectively. Kedem defined the chemi-osmotic coupling as the active transport in a narrow sense,[47] whereas Mitchel defined the chemi-osmotic and osmo-osmotic coupling as the primary and secondary active transport, respectively.[48] Thus the active transport is accomplished only by the cross-coupling of the flux of species i with that of the other species and/or the chemical reaction, and the driving force is supplied by the free energy changes of the coupled processes.

## B. ENERGY CONVERSION IN ACTIVE TRANSPORT

The active transport is a sort of energy conversion process. The energy conversion between the flux of i and the coupled chemical reaction r was investigated by Kedem and Caplan[49] and by Essig and Caplan.[50] In such a chemi-osmotic coupling system, as far as the dissipation function given by $J_iX_i + J_rX_r$ is positive, the flux $J_i$ may be negative and the up-hill transport is possible. Thus a part of the chemical energy $J_rX_r$ is converted to the energy of transport.

We may define the ratios of forces X, fluxes J, and coefficients L for species i and reaction r as x, j, and Z, respectively:

$$x = X_i/X_r \tag{41}$$

$$j = J_i/J_r \tag{42}$$

$$Z = \sqrt{L_{ii}/L_{rr}} \tag{43}$$

From Equations 33 and 35, we obtain the following equations between the flux ratio j and the force ratio x:

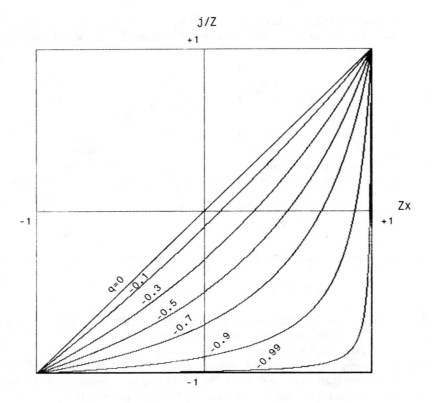

FIGURE 7. Plot of j/Z as a function of Zx for various degrees of coupling q.

$$j/Z = (q + Zx)/(1 + qZx) \qquad (44)$$

or

$$Zx = (j/Z - q)/(1 - qj/Z) \qquad (45)$$

where

$$q = L_{ir}/\sqrt{L_{ii}L_{rr}} \qquad (46)$$

The parameter q represents the degree of coupling and is in the range $-1 \leq q \leq 1$.

According to Equation 44, the flux ratio j/Z is plotted as a function of the force ratio Zx for some selected q values in Figure 7. If there is no coupling, i.e., q = 0, j/Z is directly proportional to Zx. At the extreme value of q = $-1$, Equations 44 and 45 give straight lines j/Z = $-1$ and Zx = 1, and j/Z = 1 and Zx = $-1$ at q = 1. Thus all of the transport phenomena of various degrees of coupling may be mapped in the area enclosed by these lines. Let us consider the case where Zx is positive ($X_i > 0$, $X_r > 0$). Positive coupling (q > 0) leads to an accelerated flow in the same direction, while negative coupling (q < 0) decelerates the flow and, with strong coupling, results in an evident active transport driven by the reaction r.

The force $X_i^\circ$ that leads to no apparent flow of the species i ($J_i = 0$ and therefore j/Z = 0 in Equation 44 and Figure 7) is given by the following equation and is called static head.

$$X_i^\circ = -(qX_r/Z) = -(L_{ir}/L_{ii})X_r \qquad (47)$$

Under such a condition, the difference in the chemical potential between two phases is kept constant by the coupled chemical reaction. The energy consumed in the system is given by

$$J_r X_r \ (j = 0) = L_{ii}(X_i^\circ)^2(1/q^2 - 1) \tag{48}$$

When there is no difference in the chemical potential across the membrane ($X_i = 0$ and therefore $Zx = 0$ in Equation 45 and Figure 7) and the transport is driven only by the chemical reaction, the flux $J_i^\circ$ is expressed by the following equation and is called level flow.

$$J_i^\circ = qZJ_r = (L_{ir}/L_{rr})J_r \tag{49}$$

The energy consumed in the system is given by

$$J_r X_r \ (x = 0) = (J_i^\circ)^2/L_{ii}q^2 \tag{50}$$

The efficiency of the energy conversion from chemical to transport energy is given by

$$\eta = J_i X_i / J_r X_r$$
$$= -(q + Zx)/(q + (1/Zx))$$
$$= -(q - (j/Z))/(q - (Z/j)) \tag{51}$$

With varying q values, the efficiency is plotted as a function of $Zx$ in Figure 8. For a given degree of coupling, the maximum efficiency is obtained at $Zx_{max}$:

$$Zx_{max} = -q/(1 + \sqrt{1 - q^2}) \tag{52}$$

Therefore the maximum efficiency is given by the following equation and is shown in Figure 8 (broken line).

$$\eta_{max} = q^2/(1 + \sqrt{1 - q^2})^2 \tag{53}$$

In this energy conversion system, the output $(-J_i X_i)$ is expressed as:

$$-J_i X_i = -(L_{ii}X_i^2 + L_{12}X_i X_r) \tag{54}$$

With constant $X_r$, the output is maximized when

$$-X_i/J_i = 1/L_{ii} \tag{55}$$

With constant $J_r$, the output is maximized when

$$-X_i/J_i = 1/L_{ii}(1 - q^2) \tag{56}$$

Thus for constant $X_r$ the maximal output is realized when the load resistance is equal to the internal resistance, while for constant $J_r$ the maximal output is obtained when the load conductance is equal to the internal conductance.

## C. MEMBRANE TRANSPORT MODELS

Actual active transport phenomena through liquid membrane may be treated in terms of

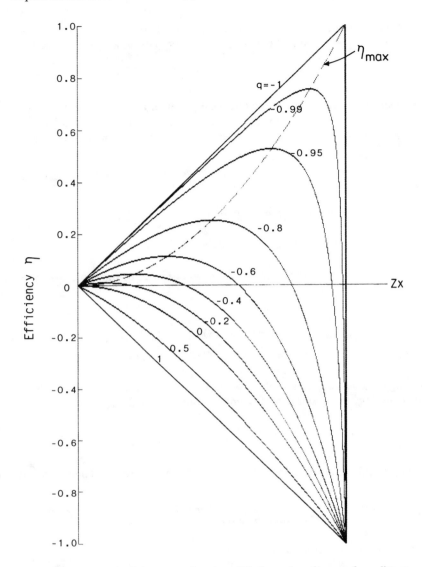

FIGURE 8.    Plot of efficiency as a function of Zx for various degrees of coupling q; maximum efficiency for each q value as a broken line.

a couple of model reaction systems discussed below. In the osmo-osmotic coupling, the flow of species i is driven by the dissymmetric distribution of the coupled species j, while in the chemi-osmotic coupling, the dissymmetric nature induced by chemical reaction r is needed in the carrier system at both interfaces.

## 1. Countertransport

The countertransport, or antiport, is one of the representative models of the osmo-osmotic coupling-driven membrane transport systems, which was mechanistically studied by Wilbrandt and Rosenberg.[51,52] A typical model system is illustrated in Figure 9,[4,6] in which a carrier X can bind either of species i or j at the same time to form a complex iX or jX, which in turn permeates through the membrane to effect transfer. The dissociation constants for the complexes iX and jX are defined as

$$K_i = C_i C_X / C_{iX} \qquad (57)$$

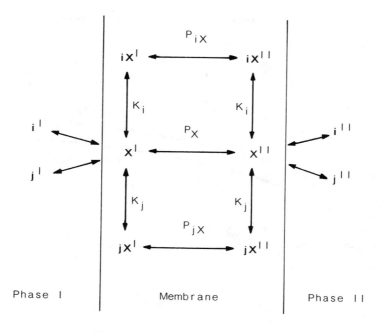

FIGURE 9. A model for countertransport.

$$K_j = C_jC_X/C_{jX} \tag{58}$$

Usually the total carrier concentration in the membrane does not change throughout the transport process, i.e.,

$$C_X^I + C_X^{II} + C_{iX}^I + C_{iX}^{II} + C_{jX}^I + C_{jX}^{II} = C_t \tag{59}$$

At the steady state,

$$P_XC_X^I - P_XC_X^{II} + P_{iX}C_{iX}^I - P_{iX}C_{iX}^{II} + P_{jX}C_{jX}^I - P_{jX}C_{jX}^{II} = 0 \tag{60}$$

where $P_X$, $P_{iX}$, and $P_{jX}$ are the permeabilities of the species X, iX, and jX, respectively. Assuming the same permeability for each species, the flux of the species i from phase I to II is given by

$$J_i = \frac{J_i^{max}C_i^I}{K_i(1 + C_j^I/K_j) + C_i^I} \tag{61}$$

It is evident from the equation that the flux of species i increases with increasing concentration of i to give the ultimate maximum flux, $J_i^{max}$, but suffers interfering effect by the species j especially at high concentrations.

When there is no j in the phase I ($C_j^I = 0$), the fluxes of species i from phase I to II and from II to I are

$$J_i^{I-II} = \frac{J_i^{max}C_i^I}{K_i + C_i^I} \tag{62}$$

$$J_i^{II-I} = \frac{J_i^{max}C_i^{II}}{K_i(1 + C_j^{II}/K_j) + C_i^{II}} \tag{63}$$

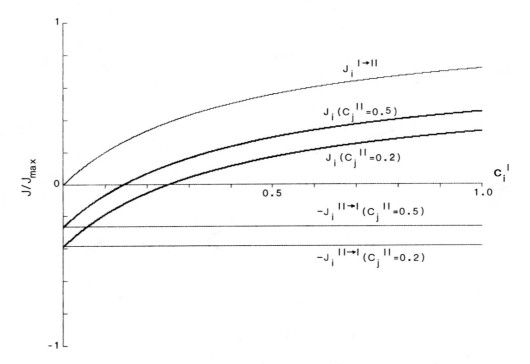

FIGURE 10.   Typical behavior of countertransport model; plots of fluxes $J_i^{I\text{-}II}$, $J_i^{II\text{-}I}$, and $J_i$ as functions of $C_i^I$ ($K_i$ = 0.4, $K_j$ = 0.2, $C_i^{II}$ = 0.5, $C_j^I$ = 0, and $C_j^{II}$ = 0.2 or 0.5).

Then the net flux of species i is given by

$$J_i = J_i^{I-II} - J_i^{II-I}$$

$$= J_i^{max}\left(\frac{C_i^I}{K_i + C_i^I} - \frac{C_i^{II}}{K_i(1 + C_j^{II}/K_j) + C_i^{II}}\right) \tag{64}$$

In Figure 10, the changes of the fluxes $J_i^{I\text{-}II}$, $J_i^{II\text{-}I}$, and $J_i$ as functions of $C_i^I$ are illustrated for the countertransport system, in which $K_i$ = 0.4, $K_j$ = 0.2, $C_i^{II}$ = 0.5, $C_j^I$ = 0, and $C_j^{II}$ = 0.2 or 0.5. The figure shows that, even if the concentrations of species i at both phases I and II are the same ($C_i^I = C_i^{II}$), an appreciable net flux from phase I to II is produced.

From Equation 64 the net flux becomes zero when

$$\frac{C_i^{II}}{C_i^I} = 1 + \frac{C_j^{II}}{K_j} \tag{65}$$

or

$$C_i^I = K_j C_i^{II}/(K_j + C_j^{II}) \tag{66}$$

Thus the transport of species i, though mechanistically active, is down-hill in the concentration range $0 < C_i^I < K_j C_i^{II}/(K_j + C_j^{II})$, while the apparent up-hill transport of species i against its concentration gradient is observed when $C_i^I$ is greater than $K_j C_i^{II}/(K_j + C_j^{II})$.

It is interesting to discuss the detailed mechanism of this up-hill transport in relation to the concept of coupling. In this case, since there is no j in phase I, the flux of i from phase

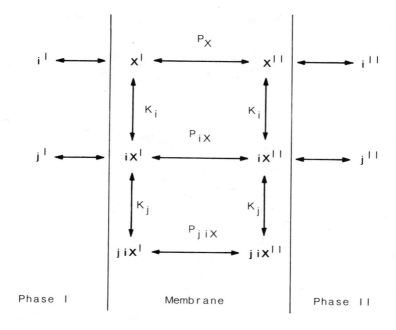

FIGURE 11.. A model for cotransport.

I to II, $J_i^{I-II}$ in Equation 62, is independent of j. However, the reverse flux, $J_i^{II-I}$, suffers interfering effect due to the presence of species j as demonstrated by Equation 63. In other words, the species j competes with i for complexation with the carrier X to diminish the reverse flux of i, which is responsible for the coupling phenomenon, and the energy necessary for the up-hill transport of species i is supplied by the down-hill transport of species j.

## 2. Cotransport

The cotransport, or symport, is another representative model of an osmo-osmotic coupling-driven membrane transport system, which was mechanistically studied by Schulz et al.[2,53] A typical model system is illustrated in Figure 11,[4,6] in which a carrier X first binds species i and then j to form complex iX and then jiX successively; no jX or ijX is produced. Similarly, the decomplexation process of the jiX proceeds through iX. The first- and second-step dissociation constants for the complex jiX are defined as:

$$K_i = C_i C_X / C_{iX} \tag{67}$$

$$K_j = C_j C_{iX} / C_{jiX} \tag{68}$$

Assuming the same permeability for the free carrier X and the complexes iX and jiX ($P_X = P_{iX} = P_{jiX} = P$), the fluxes of species i and j are represented as:

$$J_i^{I-II} = \frac{2PC_X C_i^I}{(K_i K_j / (K_j + C_j^I)) + C_i^I} \tag{69}$$

$$J_i^{II-I} = \frac{2PC_X C_i^{II}}{(K_i K_j / (K_j + C_j^{II})) + C_i^{II}} \tag{70}$$

$$J_j^{I-II} = \frac{2PC_X C_i^I C_j^I}{(K_i K_j + C_i^I (K_j + C_j^I))} = \frac{J_i^{I-II} C_j^I}{K_j + C_j^I} \tag{71}$$

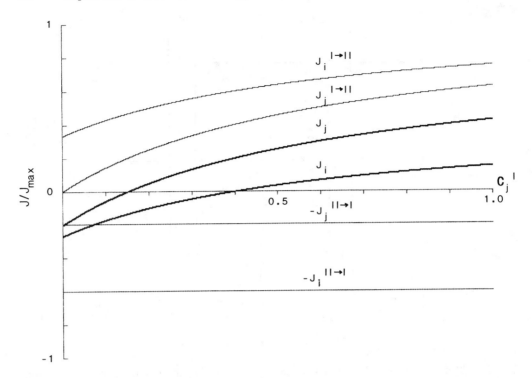

FIGURE 12.   Typical behavior of cotransport model; plots of partial fluxes $J_i^{I-II}$, $J_i^{II-I}$. $J_j^{I-II}$, $J_j^{II-I}$, and the net fluxes $J_i$ and $J_j$ as functions of $C_j^I$ ($K_i = 0.4$, $K_j = 0.2$, $C_i^I = 0.2$, $C_i^{II} = 0.4$, and $C_j^{II} = 0.1$).

$$J_j^{II-I} = \frac{2PC_xC_i^{II}C_j^{II}}{(K_iK_j + C_i^{II}(K_j + C_j^{II}))} = \frac{J_i^{II-I}C_j^{II}}{K_j + C_j^{II}} \tag{72}$$

In Figure 12, the changes of the partial fluxes $J_i^{I-II}$, $J_i^{II-I}$, $J_j^{I-II}$, $J_j^{II-I}$, and the net fluxes $J_i$ and $J_j$ are plotted as functions of $C_j^I$ for the typical cotransport system, in which $K_i = 0.4$, $K_j = 0.2$, $C_i^I = 0.2$, $C_i^{II} = 0.4$, and $C_j^{II} = 0.1$. The figure indicates that, at low concentrations of species j in phase I, the species i is transported from phase II to I according to its concentration gradient, but at high j concentrations the up-hill transport of i is observed. This is driven by the cooperative transport of species j and is energized by the down-hill transport of species j. In contrast to the countertransport in which the species i and j compete for the carrier, in the cotransport system both species bind cooperatively with the carrier to exhibit the coupling phenomenon.

According to Equations 69 and 70, the flux of species i becomes zero when

$$\frac{C_i^{II}}{C_i^I} = \frac{K_j + C_j^I}{K_j + C_j^{II}} \tag{73}$$

and, according to Equations 71 and 72, the flux of species j becomes zero when

$$\frac{C_j^{II}}{C_j^I} = \frac{C_i^I(K_i + C_i^{II})}{C_i^{II}(K_i + C_i^I)} \tag{74}$$

The degrees of coupling between the fluxes of i and j may be calculated by

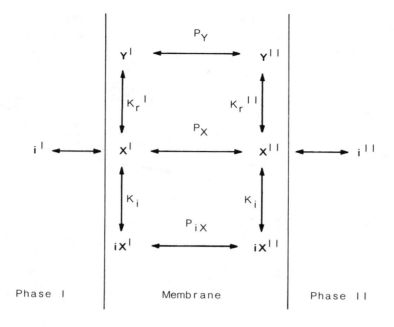

FIGURE 13.   A model for active transport driven by chemi-osmotic coupling.

$$\frac{J_j^{I-II}}{J_i^{I-II}} = \frac{C_j^I}{K_j + C_j^I} \tag{75}$$

$$\frac{J_j^{II-I}}{J_i^{II-I}} = \frac{C_j^{II}}{K_j + C_j^{II}} \tag{76}$$

These equations indicate that the degree of coupling increases as the dissociation constant $K_j$ for the complex jiX decreases or the complex becomes stable.

### 3. Chemi-Osmotic Coupling

In the active transport driven by the osmo-osmotic coupling, the energy necessary for the up-hill transport is supplied by the energy released by the coupled down-hill transport of the other species. Therefore the dissymmetric distribution of the coupling species is a prerequisite. On the other hand, the dissymmetric nature exists not in the transported species, but in the carrier system in the chemi-osmotic coupling cases, and the energy is supplied by the chemical or photochemical transformation of the carrier system.

A simple model system of chemi-osmotic coupling is shown in Figure 13.[4,6] In this system,[54] the carrier X either binds a species i to produce a complex iX or undergoes a reversible chemical reaction to produce Y, which is unable to bind the species i. Assuming the dissociation constant $K_i$ for the complex iX and the different equilibrium constants $K_r^I$ and $K_r^{II}$ for the interconversion reaction of X and Y at both interfaces of the membrane, we obtain the net flux of species i

$$J_i = J_i^{I-II} - J_i^{II-I}$$

$$= J_i^{max} \left( \frac{C_i^I}{K_i(1 + K_r^{II}) + C_i^I} - \frac{C_i^{II}}{K_i(1 + K_r^{II}) + C_i^{II}} \right) \tag{77}$$

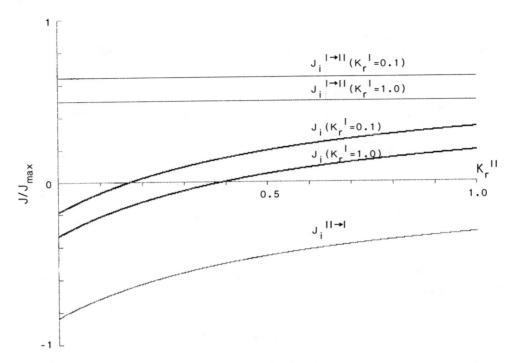

FIGURE 14.    Typical behavior of active transport driven by chemi-osmotic coupling: plot of fluxes $J_i^{I-II}$, $J_i^{II-I}$, and $J_i$ as functions of $K_r^{II}$ ($K_i = 0.1$, $K_r^I = 0.1$ or $1.0$, $C_i^I = 0.2$, and $C_i^{II} = 0.5$).

In Figure 14, the changes of the fluxes $J_i^{I-II}$, $J_i^{II-I}$, and $J_i$ as functions of $K_r^{II}$ are illustrated for the chemi-osmotic coupling transport system, in which $K_i = 0.1$, $K_r^I = 0.1$ or $1.0$, $C_i^I = 0.2$, and $C_i^{II} = 0.5$. When the equilibrium constant $K_r^{II}$ at phase II is small, the transport of i is downhill to produce negative flux, but at high $K_r^{II}$ the uphill transport of species i is observed. This is enabled by the different carrier concentrations at both sides of the membrane, which are determined by the different equilibrium constant $K_r$.

According to Equation 77, the transport of species i apparently ceases when

$$\frac{C_i^{II}}{C_i^I} = \frac{1 + K_r^{II}}{1 + K_r^I} \tag{78}$$

It is evident that, when $K_r^{II} > K_r^I$, the transport of species i is downhill in the concentration range $0 < C_i^I < (1 + K_r^I)C_i^{II}/(1 + K_r^{II})$ and uphill at $C_i^I > (1 + K_r^I)C_i^{II}/(1 + K_r^{II})$.

The actual active transport systems for electron, metal, and organic ions coupled with a flux of electron, proton, or metal ion or with chemical and/or photochemical reactions have widely been investigated by using a variety of charged or ionizable carriers including cephalin, stearic acid, quarternary ammonium ions, arenesulfonic acid, crown ether carboxylic acid, cryptands, and so on.[31,55-79]

## V. CONCLUSION

The mechanistic principles have been described for the major categories of the transmembrane phenomena: simple diffusion, passive transport, and active transport through membrane. The kinetic equations presented here may be applicable to a variety of membrane configurations, including artificial and biomembrane systems described in the other chapters of this book. The investigations of the membrane transport phenomena will be further developed for highly selective mass transfer systems and also for highly energy-efficient systems. In this context, the liquid membrane continues to serve as a model system for mechanistic investigations and, to minor extent, as practically applied supported membrane systems.

# REFERENCES

1. **Starzak, M. E.,** *The Physical Chemistry of Membranes,* Academic Press, Orlando, 1984.
2. **Schultz, S. G.,** *Basic Principles of Membrane Transport,* Cambridge University Press, 1980.
3. **Noble, R. D. and Way, J. D., Eds.,** *Liquid Membranes: Theory and Applications,* American Chemical Society, 1987.
4. **Seno, M.,** *Chemistry of Membrane (Maku no Kagaku),* Dainihon Tosho, Tokyo, 1987.
5. **Nakagaki, M.,** *Physical Chemistry of Membranes (Maku Butsuri Kagaku),* Kitami Shobo, Kyoto, 1987.
6. **Seno, M.,** Principles of active transport, in *Basic Membrane Technology (Maku Gijutsu no Kiso),* Nihon Maku Gakkai, Ed., Kitami Shobo, Kyoto, 1983, 83.
7. **Hanai, T.,** *Membrane and Ion (Maku to Ion),* Kagaku Dojin, Kyoto, 1978.
8. **Kakutani, T., Nishiwaki, Y., Osakai, T., and Senda, M.,** On the mechanism of transfer of sodium ion across the nitrobenzene/water interface facilitated by dibenzo-18-crown-6, *Bull. Chem. Soc. Jpn.,* 59, 781, 1986.
9. **Reusch, C. F. and Cussler, E. L.,** Selective membrane transport, *AIChE J.,* 19, 736, 1973.
10. **Christensen, J. J., Lamb, J. D., Izatt, S. R., Starr, S. E., Weed, G. C., Astin, M. S., Stitt, B. D., and Izatt, R. M.,** Effect of anion type on rate of facilitated transport of cations across liquid membranes via neutral macrocyclic carriers, *J. Am. Chem. Soc.,* 100, 3219, 1978.
11. **Lamb, J. D., Christensen, J. J., Oscarson, J. L., Nielsen, B. L., Asay, B. W., and Izatt, R. M.,** The relationship between complex stability constants and rates of cation transport through liquid membranes by macrocyclic carriers, *J. Am. Chem. Soc.,* 102, 6820, 1980.
12. **Lamb, J. D., Christensen, J. J., Izatt, S. R., Bedke, K., Astin, M. S., and Izatt, R. M.,** Effects of salt concentration and anion on the rate of carrier-facilitated transport of metal cations through bulk liquid membranes containing crown ethers, *J. Am. Chem. Soc.,* 102, 3399, 1980.
13. **Igawa, M., Matsumura, K., Tanaka, M., and Yamabe, T.,** Ion permselectivity of liquid membrane containing dibenzo-30-crown-10 as a carrier, *Nihon Kagaku Kaishi,* 625, 1981.
14. **Igawa, M., Matsumura, K., Tanaka, M., and Yamabe, T.,** Ion transport mechanism through liquid membrane containing crown ether, *Nihon Kagaku Kaishi,* 826, 1985.
15. **Christensen, J. J., Lamb, J. D., Brown, P. R., Oscarson, J. L., and Izatt, R. M.,** Liquid membrane separations of metal cations using macrocyclic carriers, *Sep. Sci. Technol.,* 16, 1193, 1981.
16. **McBride, D. W., Jr., Izatt, R. M., Lamb, J. D., and Christensen, J. J.,** Cation transport in liquid membranes mediated by macrocyclic crown ether and cryptand compounds, in *Inclusion Compounds,* Vol. 3, Atwood, J. L., Davies, J. E. D., and MacNicol, D. D., Eds., Academic Press, London, 1984, chap. 16.
17. **Behr, J. P., Kirch, M., and Lehn, J. M.,** Carrier-mediated transport through bulk liquid membranes: dependence of transport rates and selectivity on carrier properties in a diffusion-limited process, *J. Am. Chem. Soc.,* 107, 241, 1985.
18. **Kirch, M. and Lehn, J. M.,** Transport processes in organic chemistry. 3. Selective transport of alkali metal cations by macrobicyclic carriers through fluid membranes, *Angew. Chem.,* 87, 542, 1975.
19. **Roeske, R. W., Isaac, S., King, T. E., and Steinrauf, L. K.,** Binding of barium and calcium ions by the antibiotic beauvericin, *Biochem. Biophys. Res. Commun.,* 57, 554, 1974.
20. **Wong, K. H., Yagi, K., and Smid, J.,** Ion transport through liquid membranes facilitated by crown ethers and their polymers, *J. Membr. Biol.,* 18, 379, 1974.
21. **Lehn, J. M., Moradpour, A., and Behr, J. P.,** Antiport regulation of carrier mediated chiroselective transport through a liquid membrane, *J. Am. Chem. Soc.,* 97, 2532, 1975.
22. **Kobuke, Y., Hanji, K., Horiguchi, K., Asada, M., Nakayama, Y., and Furukawa, J.,** Macrocyclic ligands composed of tetrahydrofuran for selective transport of monovalent cations through liquid membranes, *J. Am. Chem. Soc.,* 98, 7414, 1976.
23. **Bacon, E., Jung, L., and Lehn, J. M.,** Selective transport of primary ammonium chlorides of the phenethylamine group by the macrocyclic polyether dicyclohexyl-18-crown-6, *J. Chem. Res.,* 1967, 1980.
24. **Ramdani, A. and Tarrago, G.,** Polypyrazolic macrocycles. II. A study of their complexing properties with alkali cations, *Tetrahedron,* 37, 991, 1981.

25. **Fyles, T. M., Malik-Diemer, V. A., and Whitfield, D. M.,** Membrane transport systems. II. Transport of alkali metal ions against their concentration gradients, *Can. J. Chem.,* 59, 1734, 1981.

26. **Maruyama, K., Tsukube, H., and Araki, T.,** Carrier-mediated transport of amino acid and simple organic anions by lipophilic metal complexes, *J. Am. Chem. Soc.,* 104, 5197, 1982.

27. **Shinkai, S., Kinda, H., Araragi, Y., and Manabe, O.,** Proton-driven ion transport and metal-assisted amino acid transport with an anion-capped azacrown ether, *Bull. Chem. Soc. Jpn.,* 56, 559, 1983.

28. **Izatt, R. M., Lamb, J. D., Hawkins, R. T., Brown, P. R., Izatt, S. R., and Christensen, J. J.,** Selective $M^+$-$H^+$ coupled transport of cations through a liquid membrane by macrocyclic calixarene ligands, *J. Am. Chem. Soc.,* 105, 1782, 1983.

29. **Izatt, R. M., Dearden, D. V., Brown, P. R., Bradshaw, J. S., Lamb, J. D., and Christensen, J. J.,** Cation fluxes from binary $Ag^+$-$M^{n+}$ mixtures in a water-trichloromethane-water liquid membrane system containing a series of macrocyclic ligand carriers, *J. Am. Chem. Soc.,* 105, 1785, 1983.

30. **Lamb, J. D., Brown, P. R., Christensen, J. J., Bradshaw, J. S., Garrick, D. G., and Izatt, R. M.,** Cation transport at 25°C from binary sodium ion-$M^{n+}$, cesium ion-$M^{n+}$ and strontium ion-$M^{n+}$ nitrate mixtures in a water-trichloromethane-water liquid membrane system containing a series of macrocyclic carriers, *J. Membr. Sci.,* 13, 89, 1983.

31. **Hriciga, A. and Lehn, J. M.,** pH regulation of divalent/monovalent calcium/potassium cation transport selectivity by a macrocyclic carrier molecule, *Proc. Natl. Acad. Sci. U.S.A.,* 80, 6426, 1983.

32. **Izatt, S. R., Hawkins, R. T., Christensen, J. J., and Izatt, R. M.,** Cation transport from multiple alkali cation mixtures using a liquid membrane system containing a series of calixarene carriers, *J. Am. Chem. Soc.,* 107, 63, 1985.

33. **Izatt, R. M., Izatt, S. R., McBride, D. W., Jr., Bradshaw, J. S., and Christensen, J. J.,** Cation transport at 25°C from binary $Cd^{2+}$-$M^{n+}$ mixtures in a water-trichloromethane-water liquid membrane system containing a series of macrocyclic carriers, *Isr. J. Chem.,* 25, 27, 1985.

34. **Izatt, R. M., McBride, D. W., Jr., Christensen, J. J., Bradshaw, J. S., and Clark, G. A.,** Cation transport at 25°C from binary $Tl^+$-$M^{n+}$ and $K^+$-$M^{n+}$ nitrate mixtures in a water-trichloromethane-water liquid membrane system containing a series of macrocyclic polyether carriers, *J. Membr. Sci.,* 22, 31, 1985.

35. **Izatt, R. M., Jones, M. B., Lamb, J. D., Bradshaw, J. S., and Christensen, J. J.,** Macrocycle-mediated cation transport from binary $Hg^{2+}$-$M^{n+}$ (M = metal) mixtures in a 1 *M* nitric acid-trichloromethane-1 *M* nitric acid liquid membrane system, *J. Membr. Sci.,* 26, 241, 1986.

36. **Izatt, R. M., Clark, G. A., Bradshaw, J. S., Lamb, J. D., and Christensen, J. J.,** Macrocycle-facilitated transport of ions in the liquid membrane systems, *Sep. Purif. Methods,* 15, 21, 1986.

37. **Izatt, R. M., Eberhardt, L., Clark, G. A., Bruening, R. L., Bradshaw, J. S., Cho, M. H., and Christensen, J. J.,** Macrocycle-mediated transport in a bulk 1.5 *M* nitric acid-trichloromethane-0.01 *M* $HNO_3$ membrane system of palladium (2+) and $M^{n+}$ from $Pd^{2+}$-$M^{n+}$ mixtures, *Sep. Sci. Technol.,* 22, 701, 1987.

38. **Izatt, R. M., Bruening, R. L., Wu, G., Cho, M. H., and Christensen, J. J.,** Separation of bivalent cadmium, mercury, and zinc in a neutral macrocycle-mediated emulsion liquid membrane system, *Anal. Chem.,* 59, 2405, 1987.

39. **Izatt, R. M., Lindh, G. C., Clark, G. A., Nakatsuji, Y., Bradshaw, J. S., Lamb, J. D., and Christensen, J. J.,** Protonionizable crown compounds. 5. Macrocycle-mediated proton-coupled transport of alkali metal cations in water-dichloromethane-water liquid membrane systems, *J. Membr. Sci.,* 31, 1, 1987.

40. **Izatt, R. M., Lindh, G. C., Bruening, R. L., Huszthy, P., McDaniel, C. W., Bradshaw, J. S., and Christensen, J. J.,** Separation of silver from other metal cations using pyridone and triazole macrocycles in liquid membrane systems, *Anal. Chem.,* 60, 1694, 1988.

41. **Yoshida, S. and Hayano, S.,** Kinetics of partition between aqueous solutions of salts and bulk liquid membranes containing neutral carriers, *J. Membr. Sci.,* 11, 157, 1982.

42. **Yoshida, S. and Hayano, S.,** Relationship between the extractability and the rate of transfer of potassium ion by macrocyclic carriers in liquid membrane systems, *J. Am. Chem. Soc.,* 108, 3903, 1986.

43. **Yoshida, S. and Hayano, S.,** The effect of complex stability on the rates of transport, uptake, and release of ions by macrocyclic carriers in liquid membrane systems, *J. Membr. Sci.,* 26, 99, 1986.

44. **Fujiwara, C., Inada, H., Inoue, Y., Dohno, R., and Hakushi, T.,** Transport of metal picrates through liquid membrane using crown ethers, lariat ethers, and cryptands, presented at 51st Natl. Meeting Chem. Soc. Jpn., Kanazawa, October 4-7, 1985, 277, unpublished data.

45. **Stolwijk, T. B., Sudhölter, J. R., and Reinhoudt, D. N.,** Crown ether mediated transport: a kinetic study of potassium perchlorate transport through a supported liquid membrane containing dibenzo-18-crown-6, *J. Am. Chem. Soc.,* 109, 7042, 1987.

46. **Rosenberg, T.,** On accumulation and active transport in biological systems, *Acta Chem. Scand.,* 2, 14, 1948.

47. **Kedem, O.,** Criteria of active transport, *Membr. Transp. Metab., Proc. Symp. Prague,* 87, 1961.

48. **Mitchel, P.,** Translocations through natural membranes, *Adv. Enzymol.,* 29, 33, 1967.
49. **Kedem, O. and Caplan, S. R.,** Degree of coupling and its relation to efficiency of energy conversion, *Trans. Faraday Soc.,* 61, 1897, 1965.
50. **Essig, A. and Caplan, S. R.,** Energetics of active transport processes, *Biophys. J.,* 8, 1434, 1968.
51. **Rosenberg, T. and Wilbrandt, W.,** Uphill transport induced by counterflow, *J. Gen. Physiol.,* 41, 298, 1957.
52. **Wilbrandt, W. and Rosenberg, T.,** The concept of carrier transport and its corollaries in pharmacology, *Pharmacol. Rev.,* 13, 109, 1961.
53. **Schultz, S. G. and Curran, P. F.,** Coupled transport of sodium and organic solutes, *Physiol. Rev.,* 50, 637, 1970.
54. **Rosenberg, T. and Wilbrandt, W.,** The kinetics of membrane transports involving chemical reactions, *Exp. Cell Res.,* 9, 49, 1955.
55. **Osterhout, W. J. V.,** Permeability in large plant cells and models, *Ergeb. Physiol.,* 35, 968, 1933.
56. **Rosano, H. L., Schulman, J. H., and Weisbuch, J. B.,** Mechanism of the selective flux of salts and ions through nonaqueous liquid membranes, *Ann. N.Y. Acad. Sci.,* 92, 457, 1961.
57. **Moore, J. H. and Schechter, R. S.,** Transfer of ions against their chemical potential gradient through oil membranes, *Nature,* 222, 476, 1969.
58. **Behr, J.-P. and Lehn, J.-M.,** Transport of amino acids through organic liquid membranes, *J. Am. Chem. Soc.,* 95, 6108, 1973.
59. **Schiffer, D. K., Hochhauser, A., Evans, D. F., and Cussler, E. L.,** Concentrating solutes with membranes containing carriers, *Nature,* 250, 484, 1974.
60. **Choy, E. M., Evans, D. F., and Cussler, E. L.,** A selective membrane for transporting sodium ion against its concentration gradient, *J. Am. Chem. Soc.,* 96, 7085, 1974.
61. **Sugiura, M. and Yamaguchi, T.,** Coupled transport of picrate anion through liquid membranes supported by a microporous polymer film, *J. Colloid Interface Sci.,* 96, 454, 1983.
62. **Fyles, T. M., Malik-Diemer, V. A., and Whitfield, D. M.,** Membrane transport systems. 2. Transport of alkali metal ions against their concentration gradients, *Can. J. Chem.,* 59, 1734, 1981.
63. **Fyles, T. M., Malik-Diemer, V. A., McGavin, C. A., and Whitfield, D. M.,** Membrane transport systems. 3. A mechanistic study of cation-proton coupled countertransport, *Can. J. Chem.,* 60, 2259, 1982.
64. **Fyles, T. M.,** On the rate-limiting steps in the membrane transport of cations across liquid membranes by dibenzo-18-crown-6 and lipophilic crown ether carboxylic acids, *J. Membr. Sci.,* 24, 229, 1985.
65. **Fyles, T. M.,** Interfacial surface charge and the selectivity of cation transport through a liquid membrane by a lipophilic crown ether carboxylic acid, *J. Chem. Soc., Faraday Trans. 1,* 82, 617, 1986.
66. **Castaing, M. and Lehn, J.-M.,** Efficiency, $Na^+/K^+$ selectivity and temperature dependence of ion transport through liquid membranes by $(221)C_{10}$-cryptand, an ionizable mobile carrier, *J. Membr. Biol.,* 97, 79, 1987.
67. **Izatt, R. M., Bruening, R. L., Bradshaw, J. S., Lamb, J. D., and Christensen, J. J.,** Quantitative description of macrocycle-mediated cation separations in liquid membranes and on silica gel as a function of system parameters, *Pure Appl. Chem.,* 60, 453, 1988.
68. **Shinbo, T., Kurihara, K., Kobatake, Y., and Kamo, N.,** Active transport of picrate anion through organic liquid membrane, *Nature,* 270, 277, 1977.
69. **Shinbo, T., Sugiura, M., Kamo, N., and Kobatake, Y.,** Redox reaction driven diffusion of ions through an artifical liquid membrane, *Chem. Lett.,* 1177, 1979.
70. **Shinbo, T., Sugiura, M., Kamo, N., and Kobatake, Y.,** Coupling between a redox reaction and ion transport in an artificial membrane system, *J. Membr. Sci.,* 9, 1, 1981.
71. **Shinbo, T., Sugiura, M., Kamo, N., and Kobatake, Y.,** A photoredox reaction and membrane potential in the liposome system, *Nihon Kagaku Kaishi,* 917, 1983.
72. **Yamaguchi, T., Sugiura, M., Shimakura, Y., Kamo, N, and Kobatake, Y.,** Symport of amino acids through a polymer-supported liquid membrane: evidence for diffusion-controlled transport, *Kobunshi Ronbunshu,* 43, 787, 1986.
73. **Grimaldi, J. J., Boileau, S., and Lehn, J.-M.,** Light-driven, carrier-mediated electron transfer across artificial membranes, *Nature,* 265, 229, 1977.
74. **Grimaldi, J. J. and Lehn, J.-M.,** Multicarrier transport: coupled transport of electrons and metal cations mediated by an electron carrier and a selective cation carrier, *J. Am. Chem. Soc.,* 101, 1333, 1979.
75. **Lehn, J.-M.,** Chemistry of transport processes. Design of synthetic carrier molecules, in *Physical Chemistry of Transmembrane Ion Motions,* Spach, G., Ed., Elsevier, Amsterdam, 1983, 181.
76. **Maruyama, K. and Tsukube, H.,** Metal complex mediated electron transfer system, *Chem. Lett.,* 1133, 1981.
77. **Shinkai, S., Nakaji, T., Ogawa, T., Shigematsu, K., and Manabe, O.,** Photoresponsive crown ethers. 2. Photocontrol of ion extraction and ion transport by a bis(crown ether) with a butterfly-like motion, *J. Am. Chem. Soc.,* 103, 111, 1981.

78. **Shinkai, S., Minami, T., Araragi, Y., and Manabe, O.,** Redox-switched crown ethers. 2. Redox-mediated monocrown-biscrown interconversion and its application to membrane transport, *J. Chem. Soc., Perkin Trans. 2*, 503, 1985.
79. **Shinkai, S. and Manabe, O.,** Photocontrol of ion extraction and ion transport by photofunctional crown ethers, *Top. Curr. Chem., 121 (Host Guest Complex Chem. 3)*, 67, 1984.

Chapter 6

# PRACTICAL UTILIZATION OF LIQUID MEMBRANE SYSTEMS TO DETERMINATION AND SEPARATION METHODS

**Kazuhisa Hiratani and Tomohiko Yamaguchi**

## TABLE OF CONTENTS

# I. INTRODUCTION

In recent years, increasing attention has been given to the practical availability of liquid membranes in various fields.[1] Not only selective separation and recovery of substances, which may be difficult to separate in polymeric membrane systems, but also the determination of ions and other substances with high selectivity and sensitivity could be done by using liquid membrane systems (LMSs). That is, increasing attention has been devoted to the LMSs for their use not only as tools of the separation and the condensation of valuable substances (inorganic and organic ions, neutral organic molecules, gases, etc.), but also as new analytical methods. To make good use of the features of LMSs, carriers working there and components organizing the membranes should be carefully chosen. LMSs are much more disadvantageous for membrane stability and durability (degradation and damage of membranes due to loss of membrane components), contamination of a system, complexity of apparatus, etc. than polymeric membrane systems. Action taken following substantial investigations has improved these problems, however, and practical utilization of liquid membranes is now in progress. Chemical plants of emulsion type liquid membranes have recently begun operation.[2]

In this chapter, a number of examples of carriers which work in the LMSs and their practical aspects will be described. Availability of carriers which can either detect or separate interesting and valuable target substrates in the LMSs will be focused on, but the various problems still inherent with certain types of liquid membranes and in the apparatus[1] will only be described in a few particular cases.

# II. DETERMINATION

## A. GENERAL

The utility of liquid membranes in ion analysis has become important with the successful development of artificial receptors with excellent capabilities.[3] In the last 20 years, ion-selective electrodes (ISEs) composed of liquid membrane containing neutral carrier have been improved, and some are already commercially available. Since the application of naturally occurring antibiotic, valinomycin (Structure 1) to potassium and rubidium ion-selective electrodes as an ion sensing agent has been reported,[4] various kinds of ion sensing agents for ion selective electrodes have been synthesized, and a number of ion sensing agents exhibiting excellent ion selectivity have been developed.

On the other hand, electrodes not only for the detection of organic molecules but also for the discrimination between optical isomers of organic ammonium ions and amino acid derivatives have been constituted based on the development of receptors for organic molecules. It should additionally be noted that carriers have been developed which can exhibit excellent ion-selectivity not only in electrochemical but also optical (spectroscopic) determination.

Among the features of these LMSs are miniaturization, multifunctionality, high selectivity, and high sensitivity. The compact construction of their apparatus also makes them useful in various fields. LMSs are expected to be particularly applicable to the quantitative measurement of ion concentration in a small part of a bio-cell or in blood.[5,6]

## B. INORGANIC IONS
### 1. Ion Selective Electrodes

Liquid membrane type ISEs which can exhibit selectivity for inorganic ions, especially alkali and alkaline earth metal ions, have become increasingly important for detection and quantitative analysis of these ions in biological systems. As shown in Figure 1, thin polymer matrix (e.g., PVC) film containing ion-selective sensing agents together with plasticizer is

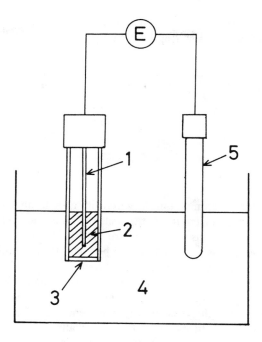

FIGURE 1. Schematic model of membrane electrode measuring circuit and cell assembly. Cell construction (an example): (1) Ag/AgCl; (2) 0.01 mol dm$^{-3}$ LiCl; (3) PVC supported liquid membrane; (4) test solution; (5) 0.1 mol dm$^{-3}$ NH$_4$NO$_3$ (or LiOAc)/sat. KCl/AgCl/Ag.

attached to the top of an electrode. The primary ion can be detected by the change of membrane potentials as an electrical response which results from the uptake of ions into the membrane and the transport through it by the sensing agent.[6]

The first successful result was gained in the early stage by electrodes containing valinomycin (Structure 1) as ion-sensing agent. This electrode exhibits a potassium ion-selectivity against sodium ion (log k$^{pot}$ (K,Na) $= -4.1$)[7] far superior to usually available glass and solid electrodes, and is now commercially available as a potassium ion-selective electrode.

Following this success, many researchers studied sensing agents for ISEs, and synthetic ionophores, e.g., crown ethers and their analogues, have been developed and tested for this purpose. A series of noncyclic diamide compounds by Simon's group, and some crown and bis(crown) compounds by Shono and Kimura's group have shown good results on ion selectivity. Diamides (Structures 2 and 3) have lithium and calcium ions selectivity, respectively.[8,9] The latter, in particular, exhibits the best selectivity among calcium ion sensing agents reported so far. It should be noted that diamide (Structure 4) can exhibit uranyl ion selectivity.[10] Thus, such noncyclic diamide compounds have been demonstrated to exhibit selective EMF response for various ions dependent upon their chain structures and/or substituents. Bis(crown) type compounds (Structures 5 and 6), on the other hand, exhibit high sodium and potassium selectivities, respectively.[11,12] 14-Crown-4 derivative (Structure 7) and cyclic tetramer of tetrahydrofuran (Structure 8) have been found to be excellent lithium ion-selective carriers.[13,14]

The noncyclic compounds, 3,3-bis(8-quinolyloxymethyl)oxetane (Structure 10)[15] and 2,9-di-n-butyl-1,10-phenanthroline (Structure 11)[16] have also exhibited excellent lithium ion selectivity.[15,16] The liquid membrane-type ISEs based on the latter compound (Structure 11) have revealed extremely good lithium ion selectivity among alkali and alkaline earth metal ions, although they are more sensitive to proton concentration than lithium ion as shown in Table 1. The calibration plots for the electrodes based on Structures 10 and 11 show a near-

## TABLE 1
### Selectivity Coefficients, Log k$^{pot}$ (Li,M), of Li$^+$-Selective Carriers

| Carrier | Log k$^{pot}$ (Li,M) | | | | | |
| --- | --- | --- | --- | --- | --- | --- |
| | H$^+$ | Na$^+$ | K$^+$ | Mg$^{2+}$ | Ca$^{2+}$ | Ref. |
| 2 | 0.95 | −2.45 | −2.6 | −4.0 | −2.9 | 8b |
| 7 | −3.1 | −2.4 | −2.3 | −4.3 | −4.7 | 13 |
| 8 | — | −2.3 | −3.0 | — | — | 14 |
| 10 | 2.2 | −2.2 | −3.0 | −1.5 | −1.1 | 15 |
| 11 | 2.4 | −3.2 | −3.5 | −3.2 | −3.3 | 16 |
| Required value | 2.1 | −4.3 | −2.8 | −3.4 | −3.6 | 8b |

Nernstian response, and their minimum detection limit for lithium ion is about $5 \times 10^{-5}$ mol dm$^{-3}$. Noticeably, the selectivity coefficient for lithium ion relative to sodium ion in the PVC membrane electrode based on Structure 11, log k$^{pot}$ (Li,Na), reaches −3.2 in the separate solution method. This electrode has the potential for use in the direct measurement of lithium concentration in a biological cell and blood.

Much attention has been paid to clinical lithium ion determination because lithium carbonate is used as an effective therapeutic agent in the treatment of manic depression,[17] and the development of a highly selective lithium sensor has been demanded to monitor lithium ion concentration in the blood.[18] Both clinically required and recently reported selectivity coefficients for lithium ion relative to other ions are listed in Table 1. Carriers with even better lithium ion selectivity are desired for use in the analysis of lithium in the blood.

Besides the ISEs described above, the liquid membrane-type pH electrode should be noticed because it has the capability, different from the pH glass electrode, to miniaturize sensing parts of the electrode. The liquid membrane electrode based on the amine compound, tridodecylamine, exhibits a pH-performance comparable to the glass electrode[19]; this is applicable to biological pH measurements.

Determination of heavy metal ions is also attractive and important; for example, besides the uranyl ion-selective carrier (Structure 4) described above, dithioamide derivatives (Structure 12) and tetraalkylthiuram mono- and di-sulfides for cadmium (II) and copper (II) ISEs, respectively, have been reported.[20,21] The selectivity and sensitivity for primary heavy metal ions should be improved for practical use.

Liquid membrane-type anion-selective electrodes have been reported using Vitamin B$_{12}$ derivative,[22] an organic tin compound such as trioctyltin chloride,[23] ammonium salts, etc. These anion sensors can exhibit anion selectivity not due to their lipophilicity, but due to the difference of their affinity with carriers. At the present stage, however, their performance is still inferior to that of the cation selective electrodes. The molecular design of anion carriers with excellent selectivity should be further investigated.

## 2. Photospectroscopic Ion Determination

Another important determination method using liquid membrane systems is the spectrophotometric method. One of the representative systems utilizes optical fiber, the tip of which is coated by liquid membrane-type thin film.[24] Although this method is not yet fully proven, irradiation of appropriate emission wavelength results in selective fluorescence behavior toward a specific ion. For example, when the optical fiber system described above containing antibiotic ionophore A2318, Structure 13, is used, the A23187-based fluorescence intensity changes with the calcium ion concentration in aqueous solution by irradiation of emission wavelength, whereas it changes very little in the presence of other alkali and alkaline earth metal ions.[25]

The influence of cations on the fluorescence behavior of carriers (Structures 10 and 11),

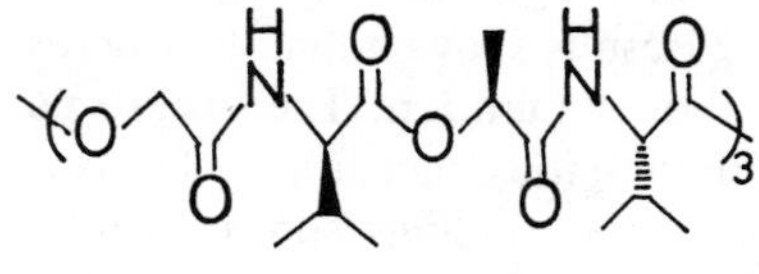

**1**

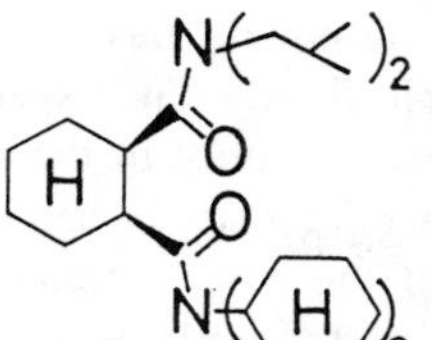

**2**

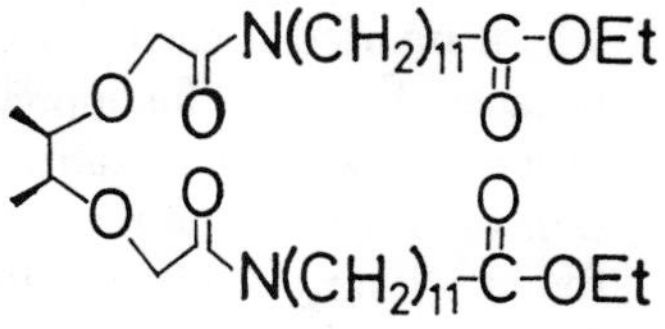

**3**

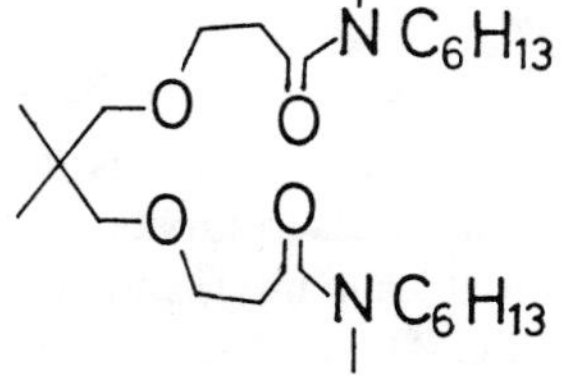

**4**

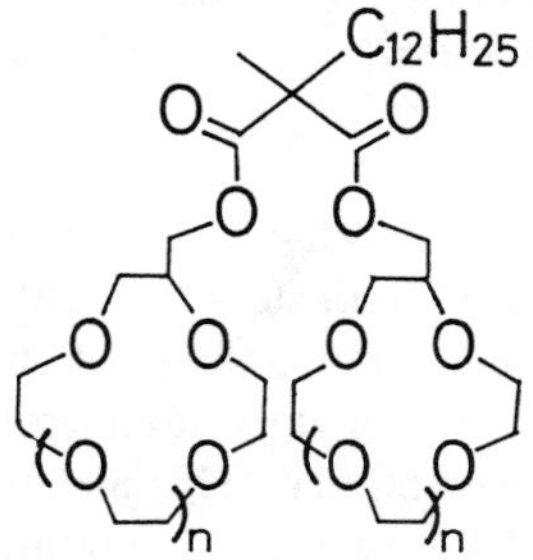

n = 1 ; **5**
n = 2 ; **6**

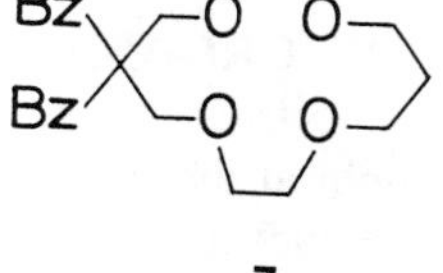

**7**

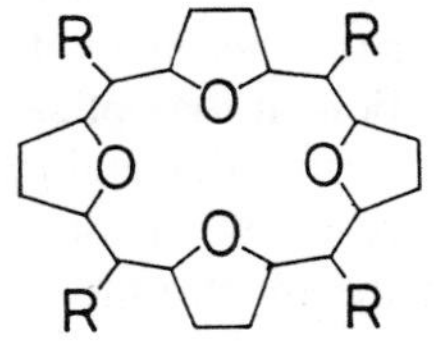

R = i–Pr ; **8**
R = Me ; **9**

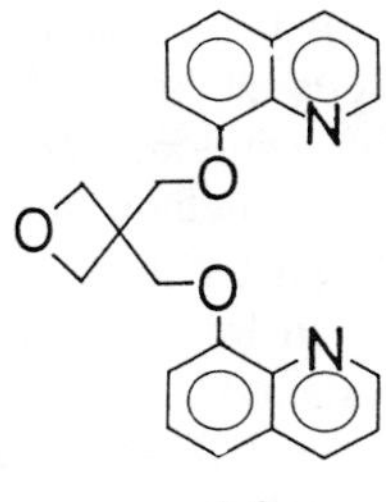

**10**

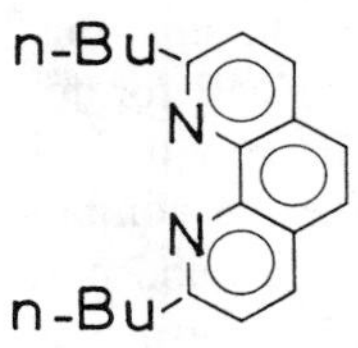

**11**

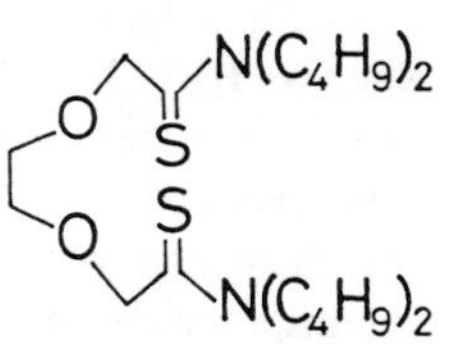

**12**

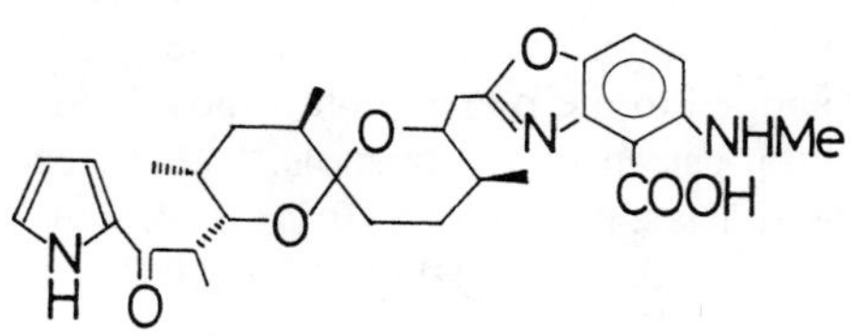

**13**

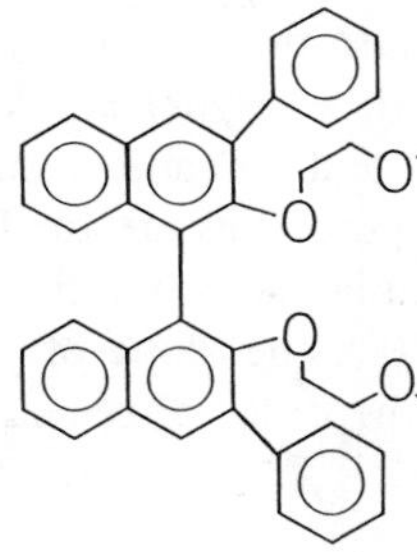

**14**

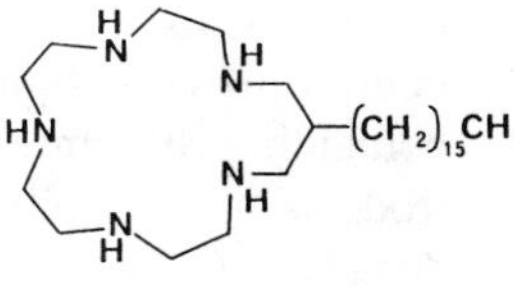

**15**

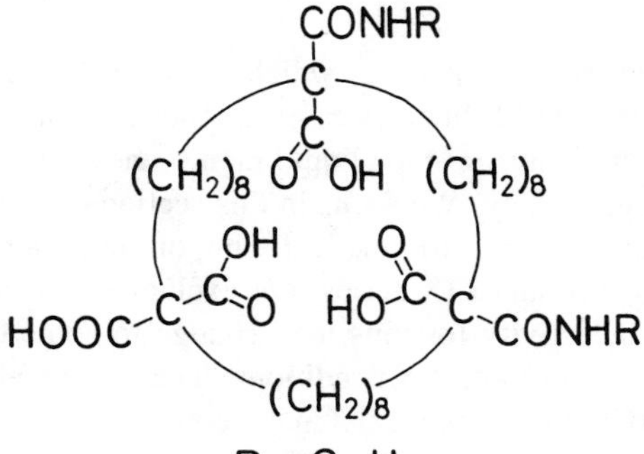

R = C₁₈H₃₇

**16**

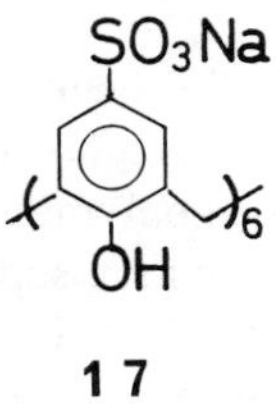

**17**

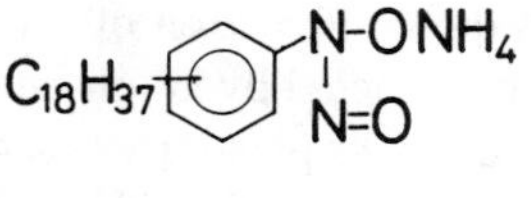

**18**

which are used as carriers for lithium ISEs, has recently been investigated in organic solvents, because they have an appreciable fluorescence spectrum.[26] Results showed that the fluorescence intensity remarkably increased only in the presence of lithium ion. The linear relationship between the fluorescence intensity of carrier and lithium ion concentration has been found within a limited range of the latter's concentration. These findings appear to make future use of liquid membrane coated optical fiber possible in ion determination.

Furthermore, various ionophores exhibiting spectroscopic cation-selectivity have been reported.[27] It is expected that such ionophores will be applicable to spectrophotometric ion determination in the LMSs from now on.

## C. ORGANIC IONS

ISEs based on chiral crown ether in PVC-plasticizer matrix are now powerful tools for chiral discrimination of amines and amino esters.[28,29] The highest chiral discrimination is achieved[30] by use of chiral crown ether possessing 1,1'-bis(3-phenyl-2-naphthyl) moiety, Structure 14.[31] The R/S selectivity for phenylglycine methyl ester (PGM) was 13.1, which corresponds to a potential difference of 66 mV. This electrode also exhibits high selectivity for esters of amino acids and low affinity for coexisting metal ions. Concentrations of both enantiomers can be finely determined by use of a pair of enantioselective electrodes with opposite selectivity.[30,32] Although this "two-electrode method" is inferior to other methods such as HPLC, GC, and NMR for determination of enantiomer concentrations if the optical purity of a sample is relatively high, it is a very facile method, and a nondestructive and *in situ* method as well, like the combination of polarimetry and UV spectroscopy. Enantioselective electrodes have a good potential to play an important role in the monitoring of enantiomeric concentrations, especially in flow systems.

There have been reports on other organic ion-selective electrodes based on PVC membranes: membrane electrodes sensitive to alkylammonium,[5] benzylammonium,[29] guanidium,[33] paraquat and diquat,[34] and dipyridinium ions[35] are prepared by use of crown ethers. Cholate-sensitive electrode is also prepared using benzyldimethylcetylammonium cholate.[36]

Macrocyclic polyamines (azacrowns) are known to form 1:1 complexes with some organic acids,[37,38] catechols,[39] and polyanions such as ATP and ADP[37,40,41] in water under neutral pH conditions. These properties are finely applied to the preparation of novel ISEs containing Structure 15, which are selective for neutral catechols and phenols.[42] This electrode is also sensitive to ATP[43]; the slope of potential response for ATP is $-14.5$ mV decade$^{-1}$ ($=$ ca. 59/4), and the order of selectivity was $ATP^{4-} > HPO_4^{2-} > ADP^{3-} > ADP^{2-}$.

# III. SEPARATION

## A. GENERAL

To experimentally evaluate the functions of carriers which work in the LMSs for ion separation, U- or H-type glass tube is often used. In other cases, supported liquid membranes utilizing microporous polymeric film and emulsion-type liquid membranes are used for their practicality and to improve separation efficiency. We focus in this section on carriers which can work in any type of liquid membrane to transport practically important target ions. Well-known chelating reagents, e.g., Kelex 100 and LIX series, are excluded here.[1]

The rate of ion transport and the selectivity for ions may change to some extent other than just with the construction of LMSs and transport conditions. That is, the ability of ion transport can be affected by anions and cations which exist in aqueous phases, the properties of lipophilic solvent or plasticizer used as liquid membranes, ion concentration, temperature, stirring rate, and so on. It should be noted that the transportability (efficiency and selectivity) of carriers is not always proportional to the extractability (which is proportional to the complex formation constant) in the carrier-mediated transport through the liquid membranes.[44]

LMSs are also promising tools for the separation of organic molecules. Not only complementarity of size and shape, but also suitable interaction between host and guest molecules should be taken into consideration in the selective separation of organic molecules.

## B. INORGANIC IONS

Separation of metal ions has usually been carried out by the methods of chelate formation (liquid-liquid extraction), metal ion adsorption using resins, and so on. Metal ions are repeatedly separated and recovered by changing the conditions in a single operation. One of the most advantageous points of separation method by LMSs is that an aqueous solution containing a large amount of ions can continuously be treated by definite amounts of carrier. Stoichiometric amounts of carriers are not needed for the separation of metal ions because carriers bring metal ions over into the liquid membranes at the boundary between the source and membrane phases; complexed metal ions with carriers then diffuse to the opposite side, metal ions are released from the membrane, and carriers repeated their task of transporting metal ions from the source phase to the receiving phase through the liquid membranes.

What kinds of metal ions are now attractive for separation? Firstly, much attention has been given to the separation of uranyl ion because it is becoming important today in connection with atomic energy resources. Uranophiles (Structures 16 and 17) which were synthesized by Tabushi's group and Shinkai's group, respectively, are potential candidates for the recovery of uranyl ion from sea water.[45,46] It has been reported that tributyl phosphate (TBP) can transport uranyl ion through supported liquid membranes.[47]

Gallium has also become one of extremely noteworthy elements in recent years because of its utility in electronic materials. Separation of gallium from a mixture of aluminum and gallium is regarded as a key technology, and hence gallium-ion separation through liquid membranes has been investigated by use of carrier (Structure 18).[48]

Attention has also been increasingly paid to carriers of trivalent rare earth metal ions. Their separation by liquid-liquid extraction has been investigated using crown ethers, and their transport through liquid membranes will become important for continuous separation. The discrimination between 17 rare earth metal ions is in particular difficult because the differences in both their ionic sizes and properties are small. A sophisticated molecular design is required to make carriers for their selective transport. It is hard in general to design selective carriers for heavy metal ions because many factors, i.e., coordination numbers, directions of coordination, ionic size, charge numbers, etc., must be considered.

Among alkali and alkaline earth metal ions, lithium ion is one of the attractive and important ions for application. Although the demand for lithium has recently increased, the locations of lithium production are limited. This makes important the technology of lithium-ion recovery and condensation from dilute solutions like sea water as is now done for uranyl ion. Although lithium is already widely utilized in various industries, large amounts of it will be necessary in the future because it is a raw material in the production of tritium which is produced by slow bombing with neutrons and is a reactant in nuclear fusion reaction. Basically, interest has been taken in lithium ion because its radius is simultaneously the smallest in its naked form and yet the largest in hydrated form; therefore, it has the greatest amount of hydration energy among all alkali metal ions.[49]

The synthesis of cyclic and noncyclic lithium ionophores has been studied since 12-crown-4 and 14-crown-4 derivatives were first reported by Pedersen as cyclic lithium ionophores,[50] and their analogues were then synthesized. More recently, 14-crown-4 derivatives having phenolic-OH (Structure 19)[51] or carboxyl group (Structure 20)[52] have been prepared for lithium ion transport through liquid membranes; these can work in supported and bulk liquid membranes as well. Macrocyclic compound (Structure 19) exhibits excellent selectivity for lithium ion (Li/Na = 20 at least).

Macrocyclic compound (Structure 9)[53] which has four tetrahydrofuran units by Kobuke and cryptand [2.1.1.][54] by Lehn exhibit lithium ion selective extractability and have a large

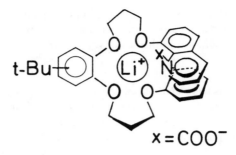

FIGURE 2.   Postulated pseudocyclic structure of Li$^+$-salt of polyether 21 on the basis of the CPK model building.

complex formation constant for lithium ion. However, the former transports sodium ion selectively over lithium ion through liquid membranes and the latter becomes inefficient in such systems. These facts mean that ion-release as well as ion-uptake is a significant step in the transport mechanism, so that a balance between ion uptake and release could be of importance in primary ion transport.[44] Thus, macrocyclic compounds are not always suitable for ion-separation through LMSs.

Noncyclic ionophores were then designed for ion-selective transport because they could easily release ions in the ion-release step. Noncyclic polyether ionophores having carboxyl group 21 can selectively transport lithium ion against its concentration gradient accompanied by proton antiport when source and receiving phases keep alkaline and acidic, respectively, in the LMSs. As shown in Figure 2, it is presumed that polyether 21 forms a pseudocyclic structure in the uptake of metal ion and releases it easily with the change of conformation at the other boundary. Polyether 21 which has two trimethylene chains is suggested able to form the pseudocavity which best fits lithium ion, based on the inspection of a CPK molecular model, spectroscopic investigation, and the results of its transport ability and that of its analogous compounds.[55,56] Simultaneous separation of lithium and potassium ions has been attempted from the source phase containing lithium, sodium, and potassium ions using a glass cell (as shown in Figure 3), where potassium and lithium ion selective ionophores (Structures 21 and 22) are contained in chloroform phases II and IV, respectively.[57] The amounts of ions transported against transport time are shown in Figure 4. Lithium and potassium ions are transported selectively into receiving phases I and V, respectively.

Noncyclic carrier (Structure 10), which can exhibit excellent lithium ion-selectivity for ion determination in the ISEs, can also transport lithium ion with excellent selectivity although its transport rate is small compared with carboxylic ionophores. The selectivities (ratio of amount of each ion transported) for lithium ion against other ions by Structure 10 reached Li/Na = 110, Li/K = 400, Li/Mg = 600, Li/Ca > 1000, and Li/Ba > 1000.[58]

Thus, neutral carrier (Structure 10) exhibits excellent lithium ion selectivity both in the determination and the separation. The X-ray structural analysis of the lithium ion complex of 1,3-bis(8-quinolyloxy)propane has been investigated. Lithium ion is pentacoordinated by two oxygen and two nitrogen atoms of the bis(quinolyloxy) compound and one oxygen atom of counter anion $ClO_4^-$.[59] Lithium ion is apparently incorporated into the pseudocyclic cavity.

Shanzer et al., on the other hand, reported that noncyclic diamide ionophore (Structure 23) transports lithium ion selectively from the inside of artificial liposomes to the outside, and the value of selectivity Li/Na reaches more than 14.[60]

One of the recent topics in ion-separation is isotopic separation of ions.[61] For the enrichment of stable isotopes, chemical systems should be much cheaper than physical separation methods such as mass separators, diffusion cells, etc. which are normally used today. Isotopic separations by the liquid-liquid extraction method in the systems where crown

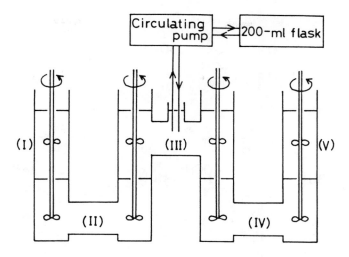

FIGURE 3. Apparatus for simultaneous separation of multiple cation species. Initial transport conditions (25°C): receiving phase (I): 15 ml of 0.05 $M$ (mol dm$^{-3}$) $H_2SO_4$ aqueous solution; chloroform liquid membrane (II): 30 ml of chloroform containing $1.5 \times 10^{-4}$ mol of 22; source phase (III): 230 ml of 0.1 $M$ LiOH + 0.1 $M$ NaOH + 0.1 $M$ KOH + 0.1 $M$ $H_2SO_4$ aqueous solution; chloroform liquid membrane (IV): 30 ml of chloroform containing $1.5 \times 10^{-4}$ mol of 21; receiving phase (V): 15 ml of 0.05 $M$ $H_2SO_4$ aqueous solution. (From Hiratani, K. et al., *J. Membrane Sci.*, 35, 91, 1987. With permission.)

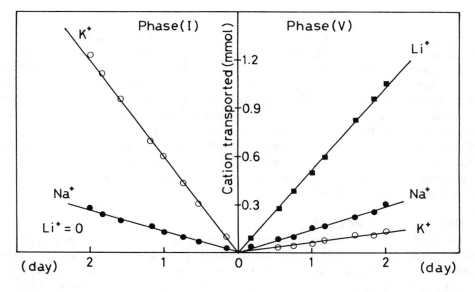

FIGURE 4. Plots of the amounts of cation transport from the source phase (III) to the receiving phases (I) and (V) by carriers 22 and 21, respectively, vs. time. (From Hiratani, K. et al., J. Membrane Sci., 35, 91, 1987. With permission.)

ethers or cryptands are used have been carried out with alkali and alkaline earth metal ions in particular.[62] Separation of $^6$Li from $^7$Li is of importance because $^6$Li is used for the production of tritium.[63] Today the separation is conducted by Hg-amalgam methods, by which slightly enriched $^6$Li (separation factor $\alpha$ = ([$^6$Li]/[$^7$Li]) $_{(org)}$/([$^6$Li]/$^7$Li]) $_{(aq)}$ = 1.05) is obtainable within one equilibrium stage[64]; a cascade method must therefore be applied. This is disadvantageous, however, because of the need for large amounts of Hg, high

**1 9**                          **2 0**

**2 1**                          **2 2**

**2 3**

solubility of Hg-amalgam into aqueous solution, and so on. While stoichiometric amounts of extractants are needed for the liquid-liquid ⁶Li extraction method, a solution containing a large quantity of lithium ion could be treated by ion-transport separation through LMSs using a comparatively small amount of carrier. Recently, isotopic separation of Li using noncyclic polyether (Structure 10) in the LMSs has been carried out, and separation factor $\alpha = 1.025$ at 25°C could be obtained.[65] This value is almost comparable to the separation factors obtained in the extraction systems using crown ethers and hence such noncyclic polyethers could be available for isotopic separation in the LMSs. In addition, isotopic separation for $Ca^{2+}$, $Na^+$, and so on has been reported by the extraction method. The development of carriers for selective and efficient separation of such isotopic ions in the LMSs is anticipated in future.

## C. ORGANIC MOLECULES

Different from inorganic ions, organic molecules are often obtained as a mixture of stereoisomers. Racemates for example are commonly found in drugs and chemicals; recent research[66] revealed that almost one-half of the medicines and 20% of the pesticides commercially available in Japan have asymmetric structures. Among these asymmetric chemicals, 40% of medicines and 90% (!) of pesticides are used as racemates. As the physical and chemical properties of isomers are generally very close to each other, sometimes required for their separation are energy-consuming processes (e.g., for mutual separation of xylenes) or tedious processes (e.g., for diastereomeric separation of racemates).

LMSs present here a facile, hopeful, and favorable method for the purpose of large-scale resolution of isomers, especially of enantiomers, and hence interest has gradually moved from laboratory curiosity to practical utility. We stress in this section the selective transport of enantiomers through LMSs. LMSs available for resolution of racemates can be classified into two general categories: those containing carrier molecules (separator system) and those combined with an enzymatic reaction process (reactor system). The former separator system is of our main concern; for the latter reactor systems, see the literature.[67-70]

The most intensively studied LMS is that for resolution of racemic amino acids. The idea was proposed as early as 1971.[71,72] The first artificial transport system of amino acids was reported by Behr and Lehn in 1973[73]; it was a bulk toluene membrane containing dinonylnaphthalene sulfonate or tetraalkylammonium salt, Aliquat 336. The first and most elegant method of enantiomeric resolution of amino esters appeared in 1974. Cram and his co-workers reported bulk LMSs containing chiral crown ethers (Structure 24, for example); an enantiomeric resolution ratio of 10 for PGM was achieved.[74] An enantiomer-resolving machine, a W-shaped LMS based on a pair of chiral crown ethers of mirror image, was also proposed.[75] A series of intensive studies on chiral crown ethers by Cram's group[76] has confirmed the efficiency of binaphthyl moiety as a chiral center.

Since that earlier period, a large number of chiral crown ethers have been synthesized,[77,78] and their specificities have been examined by use of bulk LMSs. Recent crown ethers (Structures 25-27) reported by Yamamoto et al. are comparable with binaphthyl crown ethers in their ability for chiral discrimination. They are incorporating synthetic chiral elements such as helicene,[79] twisted ethylene,[80] and biphenanthrene[22] in their ring structures, and observed resolution ratios are 7.7, 4.1, and 1.6, respectively, for PGM. Even noncyclic polyether (Structure 28) by Naemura et al. exhibits surprisingly high chiral discrimination for 1,2-diphenylethylamine with a resolution ratio of 11.5.[82]

From a practical point of view, supported liquid membranes (SLMs) have been investigated by Sugiura and his co-workers for the purpose of use in enantiomeric resolution of amino acids. Co-transport of free amino acids and $PF_6^-$ is carried out through liquid membranes of o-nitrophenylphenyl ether(NPPE), o-nitrophenyl octyl ether(NPOE), or dihalogenoalkane containing trialkylphosphate; the liquid membranes are supported by a microporous polypropylene film (Duragard 2400, equivalent to Celgard 2400) or a PVC matrix.[83] This is the first finding of transport of free amino acids through SLMs.

In place of trialkylphosphate, chiral crown ethers have been examined in supported NPPE. To date chiral crown ether 14[31] has proven best as a carrier of enantioselective transport of free amino acids through SLMs.[84,85] Yamaguchi et al. evaluated the ability of chiral crown ethers by comparison of the following parameters based on transport mechanism[86,87]: permeation coefficient P, extraction constant $K_{ext}$, and enantiomeric resolution ratio (Table 2).[85] Observed values of enantiomeric resolution ratio are 22.7 for phenylglycine, 7.2 for phenylalanine, 13.5 for leucine, and 11.2 for methionine, and others.[88]

Experimental limitations are sometimes encountered in enantioselective transport, especially with the determination of enantiomer concentration. The most promising method is HPLC, and good prepacked columns are now commercially available.[89]

Miscellaneous carriers other than crown ethers have been designed and examined for selective transport of various organic compounds. Rebek et al. transported amino acids in *zwitter-ionic* forms using center-convergent noncyclic host molecules (Structure 29).[90] Two host molecules and one amino acid make a stable complex by hydrophobic and electrostatic interaction. Nucleosides such as adenosine and deoxyadenosine can be transported as well by analogous center-convergent carriers.[91] Reversible extraction of amino acids in *nonionic* form is also achieved by Aoyama et al. by use of rhodium(III)-porphyrin derivative,[92] by which amino acid could be transported through liquid membranes.

Saccharides are highly hydrophilic molecules and hence generally insoluble in organic phases such as chloroform and dichloroethane. With the aid of proper lipophilic carriers, however, even saccharides can be transported through organic liquid membranes. 1,2- and 1,3-diols are known to react with boric acid under alkaline condition to form boric esters; the esters are decomposed under acidic condition. Using this reversible reaction, Shinbo et al. reported the transport of hexoses against their concentration gradients through bulk LMSs

**TABLE 2**

**Permeation Coefficient, P, and Extraction Constant, $K_{ext}$, and Enantiomeric Resolution Ratio, @, for Chiral Crown Ethers in Enantioselective Co-Transport of Phenylglycine Perchlorate Through Supported Liquid Membranes[87]**

| Crown ether | P[a] | $K_{ext}$ [b] | @ | Selectivity |
|---|---|---|---|---|
| (SS)-D(OEOEO)$_2$D[c] | — | — | — | — |
| (R)-D(OEOEO)$_2$F[c] | 7.8 | 1.9 | 1.5 | L |
| (R)-14 | 8.7 | 5.1 | 15.5 | D |
| (P)-25 | 1.8 | 2.0 | 1.2 | D |
| (S)-27 | 6.2 | 12.2 | 1.9 | D |
| Dibenzo-18-crown-6 | 7.8 | 4.7 | 1.0 | — |

[a]   In $10^{-6}$ cms$^{-1}$.
[b]   In $10^2 M^{-2}$ ($M$ = mol dm$^{-3}$).
[c]   D = 1,1′-binaphthyl unit, E = ethylene unit, O = oxygen. (SS)-D(OEOEO)$_2$D is analogous to crown ether 24.

containing phenylboric acid and trioctylmethylammonium chloride (TOMA).[93] The source phase must be alkaline (higher than pKa of the boric acid) and the receiving phase must be acidic (lower than the pKa). The order of transport rate is fructose = mannose > galactose > glucose. Similarly, mono- and di-saccharides as well as nucleosides are transported through SLMs containing TOMA and alkylphenyl boric acids.[94] The characteristic of these sugar-borate systems is that apparent transportees are neutral species in aqueous solutions before and after transport. The actual species transported through organic liquid membranes, however, are anionic borate esters of sugars. Coupling of the flows of two ionic species, the borate esters and another ion (Cl$^-$?), can easily take place. As a result, uphill transport of even neutral sugars is achieved against their concentration gradients.

Other candidates for the carrier of sugar-transport are macrocyclic products of resolucinol-aldehyde condensation. These macrocycles[95,96] were found some 50 years ago, and their property to form strong ion-complexes with organic amines in alkaline aqueous atmosphere was reported.[97] It was only very recently found by Aoyama et al. that macrocycle (Structure 30) extracts monosaccharides, vitamin B$_2$, and vitamin B$_{12}$ from aqueous phases into CCl$_4$ or benzene,[98] the host-guest complex is stabilized through many hydrogen bonds, and the complex is neutral. Affinity towards monosaccharides and nucleosides is determined as follows: deoxyribose, ribose > arabinose, erythrose > glucose, mannose > xylose, lyxose. Selective transport of sugars through organic liquid membranes has been carried out.

Mandelic acid and tartaric acid were historically the first targets for enantioselective transport. The possibility was suggested in 1971, based on the results of diffusion experiments of individual enantiomers through liquid membranes.[72] It is noteworthy that cyclodextrin was also examined as carrier, though the experiment was unsuccessful. Slight success of enantioselective transport of mandelate anion is based on ion-pair extraction.[99] Mandelate anion and Cl$^-$ or propionate anion are counter-transported by chiral amine (N-(1-naphthyl)methylalpha-methylbenzylamine) through bulk chloroform membranes to give a value of enantiomeric resolution ratio below 1.1.

Cyclophanes and cyclodextrins are good carriers for aromatic compounds in aqueous liquid membranes. Water-soluble cyclophane (Structure 31) is used as a carrier of lipophilic arenes through aqueous bulk LMSs.[100] Naphthalene is separated from phenanthrene by macrobicycle (Structure 32).[101] Active transport of neutral aromatic compounds driven by redox reaction might be possible with Structure 33.[102] A good review is available of recent results on complexation of neutral compounds with cyclophanes in an aqueous environment,

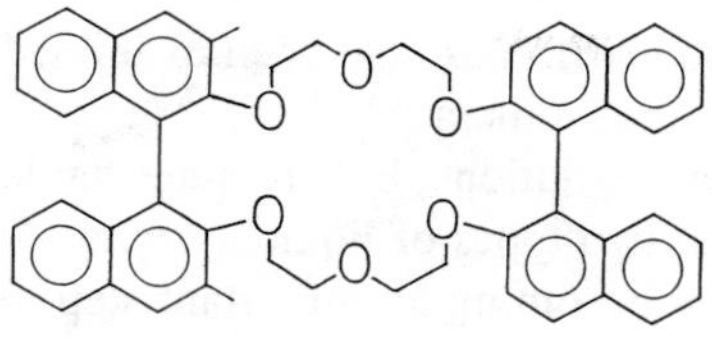

**24**

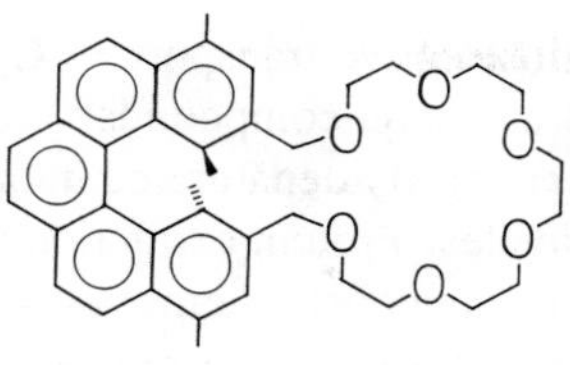

**25**

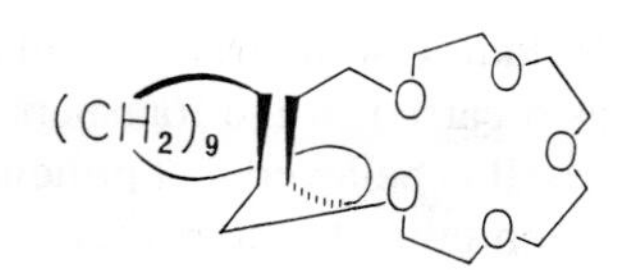

**26**

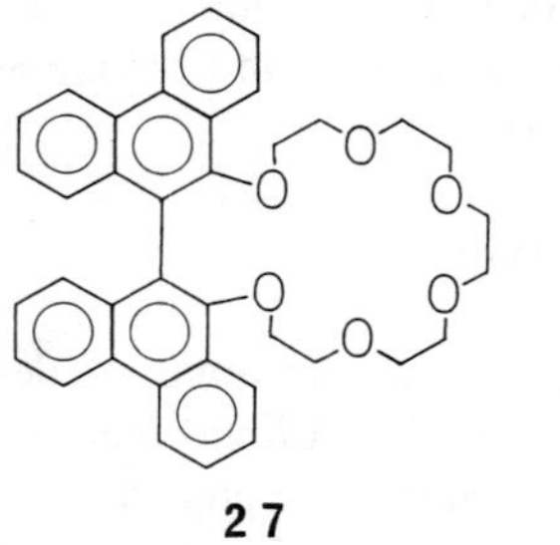

**27**

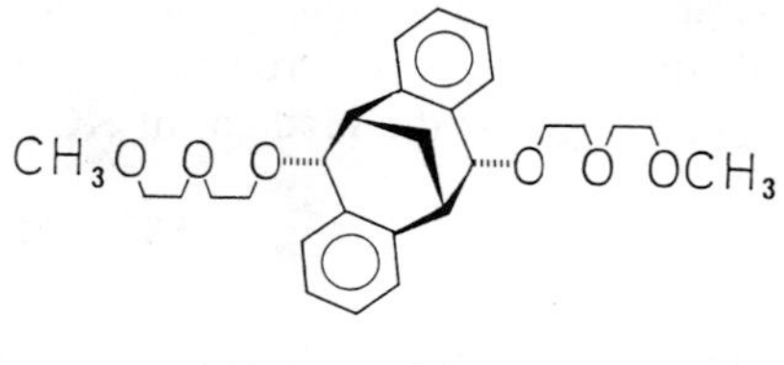

**28**

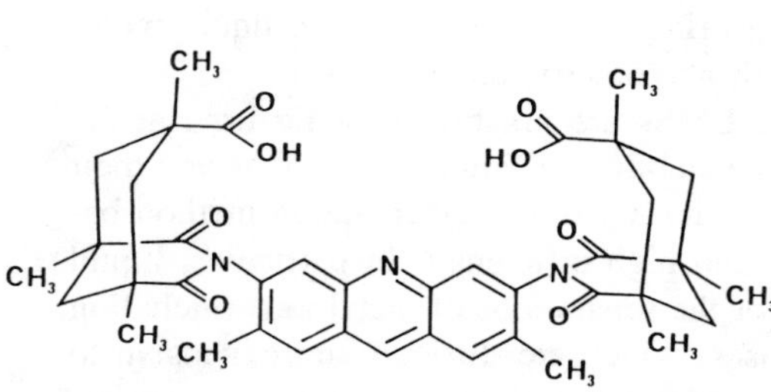

**29**

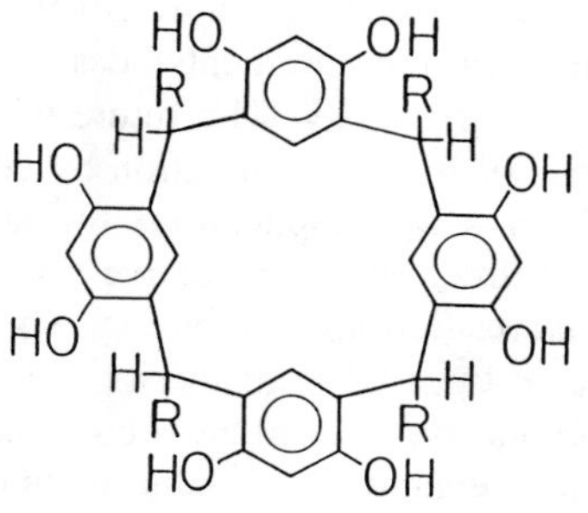

**30** $(R = C_{11}H_{23})$

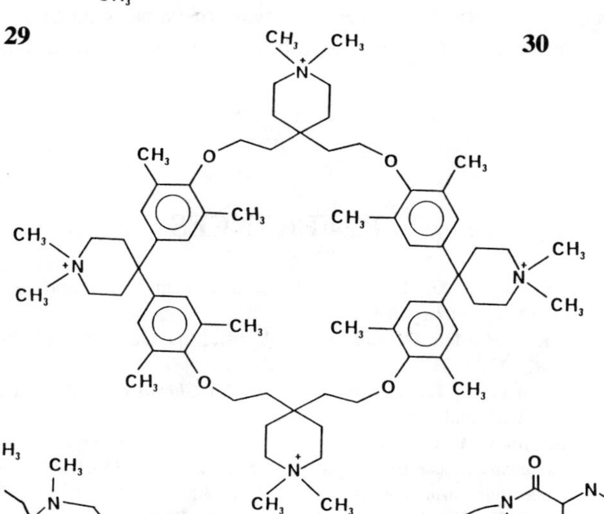

**31**

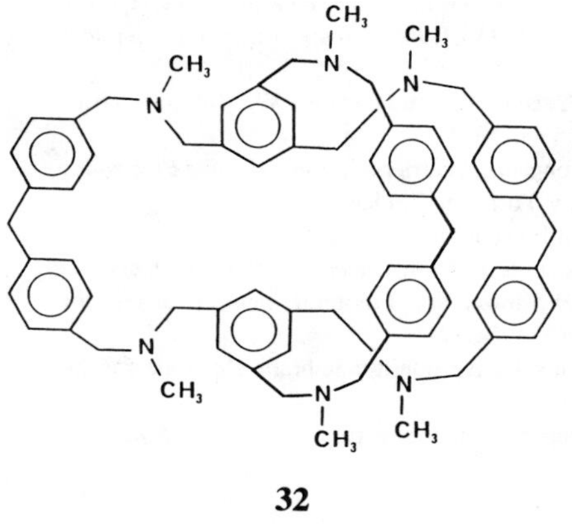

**32**

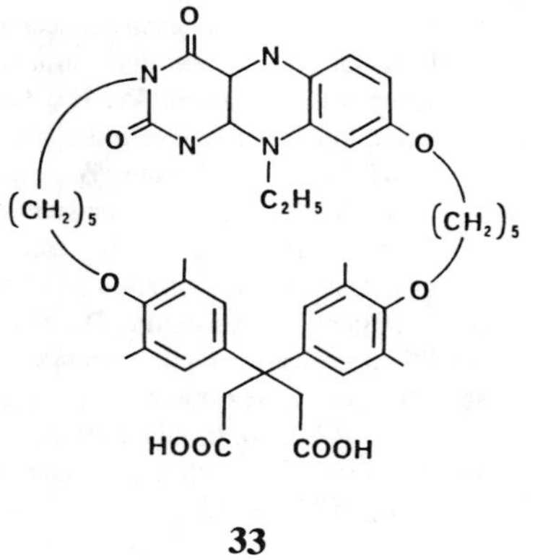

**33**

including some trials on enantioselective transport.[103] Cyclodextrin has also been examined as a carrier in the transport of aromatic compounds[104] and enantiomers.[105]

It is believed that proteins are easily denatured in organic solutions, but are quite stable when protected in reversed micelles. Protein extraction with two types of liquid systems — reversed micelles and two-phase aqueous systems — is now becoming an important step in biotechnology.[106-108] Recent attempts to utilize reversed micelles for separation and purification of proteins have progressed due to the application of protein separation through organic liquid membranes.[109] Through liquid membranes of hexane containing aerosol OT (AOT), four kinds of proteins (cytochrome c, myoglobin, lysozyme, and bovine serum (albumin) are transported against the gradient of KCl concentration.

## IV. OUTLOOK

The most successful application of LMSs to date in the areas of determination and separation is ion-determination by ion-selective electrodes. Liquid-membrane-type ion-selective electrodes have now developed so that they enjoy wide applications in medical and biochemical measurements because of their advantages of miniaturization, high selectivity, and high sensitivity. The future will see new measuring systems advanced in liquid membranes, such as determination by combination with photometric analysis.

Only a few applications for separation using LMSs are practically being used at the present time. However, active investigations on the stability of liquid membranes and their systematization are continuing. These membranes offer a promising separation method because of their inherent merits of high selectivity and high efficiency. In particular, liquid membrane systems might become advantageous for the separations of metal ions (including isotopic separation), organic compounds, and gases, which are difficult and expensive to achieve by other methods. In this sense, synthesis of carriers having other excellent capabilities and the improvement of liquid membrane systems from an engineering aspect are eagerly awaited.

## REFERENCES

1. **Noble, R. D. and Way, J. D., Eds.,** Liquid Membranes, in *ACS Symposium Series* 347, Am. Chem. Soc., Washington, D.C., 1987.
2. **Li, N. N., Chan, R. P., Naden, D., and Lai, R. W. M.,** Liquid membrane processes for copper extraction, *Hydrometallurgy*, 9, 277, 1983.
3. **Vögtle, F. and Weber, E., Eds.,** *Host-Guest Complex Chemistry*, Vol. I, II and III, Springer-Verlag, Berlin, 1981(I), 1982(II), and 1984(III).
4. **Stefanac, Z. and Simon, W.,** *In-vitro*-Verhalten von Makro-tetroliden in Membranen als Grundlage für hochselective kationenspezifische Electrodensysteme, *Chimia*, 20, 436, 1966; Ion specific electrochemical behavior of macrotetrolides in membranes, *Microchem. J.*, 12, 125, 1967.
5. **Ammann, D., Morf, W. E., Anker, P., Meier, P. C., Pretsch, E., and Simon, W.,** Neutral carrier based ion-selective electrodes, *Ion-Selective Electrode Rev.*, 5, 3, 1983.
6. **Morf, W. E. and Simon, W.,** Ion-selective electrodes based on neutral carriers, in *Ion-Selective Electrodes in Analytical Chemistry*, Freiser, H., Ed., Plenum Press, New York, 1987, chap. 3.
7. **Osswald, H. F., Asper, R., Dimai, W., and Simon, W.,** On-line continuous potentiometric measurement of potassium concentration in whole blood during open-heart surgery, *Clin. Chem.*, 25, 39, 1979.
8a. **Metzger, E., Ammann, D., Schefer, U., Pretsch, E., and Simon, W.,** Lipophilic neutral carriers for lithium selective liquid membrane electrodes, *Chimia*, 38, 440, 1984;
8b. **Metzger, E. Ammann, D., Asper, R., and Simon, W.,** Ion selective liquid membrane electrode for the assay of lithium in blood serum, *Anal. Chem.*, 58, 132, 1986.
9a. **Ammann, D., Pretsch, E., and Simon, W.,** A synthetic electrically neutral carrier for $Ca^{2+}$, *Tetrahedron Lett.*, 1972, 2473;

9b. **Simon, W., Ammann, D., Oehme, M., and Morf, W. E.,** Calcium-selective electrodes, *Ann. N. Y. Acad. Sci.,* 307, 52, 1978.

10. **Senkyr, J., Ammann, D., Meier, P. C., Morf, W. E., Pretsh, E., and Simon, W.,** Uranyl ion selective electrode based on a new synthetic neutral carrier, *Anal. Chem.,* 51, 786, 1979.

11. **Shono, T., Okahara, M., Ikeda, I., Kimura, K., and Tamura, H.,** Sodium-selective PVC membrane electrodes based on Bis(12-crown-4)s, *J. Electroanal. Chem.,* 132, 99, 1982.

12. **Kimura, K., Tamura, H., and Shono, T.,** A highly selective ionophore for potassium ions: a lipophilic bis(15-crown-5) derivative, *J. Chem. Soc., Chem. Commun.,* 1983, 492.

13. **Kimura, K., Yano, H., Kitazawa, S., and Shono, T.,** Synthesis and selectivity for Lithium of Lipophilic 14-crown-4 derivatives bearing bulky substituents or an additional binding site in the side arm, *J. Chem. Soc., Perkin Trans. II,* 1986, 1945.

14. **Tohda, K., Sasakura, H., Suzuki, K., and Shirai, T.,** presented at Chem. Soc. Japan, 54th Annual Meeting, Tokyo, April 1 to 4, 1987, 473.

15. **Hiratani, K., Okada, T., and Sugihara, H.,** 1,3-Bis(8-quinolyloxy)propane derivatives as neutral carriers for lithium ion selective electrodes, *Anal. Chem.,* 59, 766, 1987; **Okada, T., Hiratani, K., and Sugihara, H.,** Improvement of characteristics of PVC membrane lithium-selective electrodes based on noncyclic neutral carriers by the use of lipophilic additives, *Analyst,* 112, 587, 1987.

16. **Sugihara, H., Okada, T., and Hiratani, K.,** Lithium ion-selective electrodes based on 1,10-phenanthroline derivatives, *Chem. Lett.,* 1987, 2391.

17. **Tosteson, D. C.,** Lithium and mania, *Sci. Am.,* 244, 164, 1981.

18. **Gadzekpo, V. P. Y., Moody, G. J., Thomas, J. D. R., and Cristian, G. D.,** Lithium ion-selective electrodes, *Ion-Selective Electrode Rev.,* 8, 173, 1986.

19. **Schulthess, P., Shijo, Y., Pham, H. V., Pretsch, E., Ammann, D., and Simon, W.,** A hydrogen ion-selective liquid-membrane electrode based on tri-n-dodecylamine as neutral carrier, *Anal. Chim. Acta,* 131, 111, 1981; **Ammann, D., Lanter, F., Stenier, R. A., Schulthess, P., Shijo, Y., and Simon, W.,** Neutral carrier based hydrogen ion selective microelectrode for extra- and intracellular studies, *Anal. Chem.,* 53, 2267, 1981.

20. **Schneider, J. K., Hofstetter, P., Pretsch, E., Ammann, D., and Simon, W.,** N,N,N',N'-Tetrabutyl-3,6-dioxaoctan-dithioamid, Ionophor mit Selectivitat fur $Cd^{2+}$, *Helv. Chim. Acta,* 63, 217, 1980.

21. **Kamata, S., Bahle, A., and Uda, T.,** Thiuram monosulfides as a neutral carrier for copper(II)-selective membrane electrode, *Chem. Lett.,* 1988, 1247.

22. **Schulthess, P., Ammann, D., Simon, W., Christian, C., Stepanek, R., and Krautler, B.,** A lipophilic derivative of Vitamin $B_{12}$ as a selective carrier for anions, *Helv. Chim. Acta,* 67, 1026, 1984.

23. **Oesch, U., Ammann, D., Pham, H. V., Wuthier, U., Zund, R., and Simon, W.,** Design of anion-selective membranes for clinically relevant sensors, *J. Chem. Soc., Faraday Trans. I.,* 82, 1179, 1986.

24. Bioanalytical applications of fiber-optic chemical sensors, *Anal. Chem.,* 58, 766A, 1986.

25. **Suzuki, K. and Shirai, T.,** unpublished data, 1987.

26. **Hiratani, K.,** Drastic change in fluorescence intensity of noncyclic polyethers by addition of lithium ion, *J. Chem. Soc., Chem. Commun.,* 1987, 960; **Hiratani, K.,** Fluorescence spectrophotometric determination of lithium ion based on 1,3-bis(8-quinolyloxy)propane derivatives, *Analyst,* 113, 1065, 1988.

27. **Löhr, H.-G. and Vögtle, F.,** Chromo-and fluoroionophores. A new class of dye reagents, *Acc. Chem. Res.,* 18, 65, 1985.

28. **Simon, W.,** Selective transport processes in artificial membranes, in *Advances in Chemical Physics,* Vol. 39, Lefever, R. and Goldbeter, A., Eds., John Wiley and Sons, New York, 1978, 287.

29. **Bussman, W., Lehn, J.-M., Oesch, U., Plumere, P., and Simon, W.,** Enantiomer-selectivity for phenylethylammonium ion of membranes based on a chiral macrocyclic polyether, *Helv. Chim. Acta,* 64, 657, 1981.

30. **Shinbo, T., Yamaguchi, T., Nishimura, K., Kikkawa, M., and Sugiura, M.,** Enantiomer-selective membrane electrode for amino acid methyl esters, *Anal. Chim. Acta,* 193, 367, 1987.

31. **Lingenfelter, D. S., Helgeson, R. C., and Cram, D. J.,** Host-guest complexation. 23. High chiral recognition of amino acid and ester guests by host containing one chiral element, *J. Org. Chem.,* 46, 393, 1981.

32. **Bussmann, W., Morf, W. E., Vigneron, J. P., Lehn, J.-M., and Simon, W.,** Messkette zur direkten potentiometrischen Bestimmung des Enantiomerenüberschusses von 1-Phenyläthyl-ammonium-ionen, *Helv. Chim. Acta,* 67, 1439, 1984.

33. **Assubaie, F. N., Moody, G. J., and Thomas, J. D. R.,** Guanidium ion-selective electrodes based on dibenzo-27-crown-9 and tetraphenylborate, *Analyst,* 113, 61, 1988.

34. **Moody, G. J., Owusu, R. K., and Thomas, J. D. R.,** Liquid membrane ion-selective electrodes for diquat and paraquat, *Analyst,* 112, 121, 1987.

35. **Moody, G. J., Owusu, R. K., and Thomas, J. D. R.,** Studies on crown ether based potentiometric sensors for 4,4'-dipyridinium and related dications, *Analyst,* 113, 65, 1988.

36. **Campanella, L., Mazzei, F., Tomassetti, M., and Sbrilli, R.,** Polymeric membrane cholate-selective electrode, *Analyst,* 113, 325, 1988.

37. **Dietrich, B., Hosseini, M. W., Lehn, J.-M., and Sessions, R. B.,** Anion receptor molecules. Synthesis and anion-binding properties of polyammonium macrocycles, *J. Am. Chem. Soc.,* 103, 1282, 1981.

38. **Kimura, E., Sakonaka, A., Yatsunami, T., and Kodama, M.,** Macromonocyclic polyamines as specific receptors for tri-carboxylate-cycle anions, *J. Am. Chem. Soc.,* 103, 3041, 1981.

39. **Kimura, E., Watanabe, A., and Kodama, M.,** A catechol receptor model by macrocyclic polyamines, *J. Am. Chem. Soc.,* 105, 2063, 1983.

40. **Hosseini, M. W., Lehn, J.-M., and Mertes, M. P.,** Efficient molecular catalysis of ATP-hydrolysis by protonated macrocyclic polyamines, *Helv. Chim. Acta,* 66, 2454, 1983.

41. **Hosseini, M. W. and Lehn, J.-M.,** Anion receptor molecules. Chain length dependent selective binding of organic and biological dicarboxylate anions by ditopic polyammonium macrocycles, *J. Am. Chem. Soc.,* 104, 3525, 1982.

42. **Umezawa, Y., Kimura, E., et al.,** unpublished results, 1988.

43. **Umezawa, Y., Kimura, E., et al.,** unpublished results, 1988.

44. **Lehn, J.-M.,** Macrocyclic receptor molecules, *Pure Appl. Chem.,* 52, 2303, 1980; **McBride, Jr., D. W., Izatt, R. M., Lamb, J. D., and Cristensen, J. J.,** *Inclusion Compounds,* Vol. 3, Atwood, J. L., Davies, J. E. D., and MacNicol, D. D., Eds., Academic Press, London, 1984; 257; **Behr, J.-P., Kirch, M., and Lehn, J.-M.,** Carrier-mediated transport through bulk liquid membranes: Dependence of transport rates and selectivity on carrier properties in a diffusion-limited process, *J. Am. Chem. Soc.,* 107, 241, 1985.

45. **Kobuke, Y., Tabushi, I., and Kohki, O.,** Active transport of uranyl ion by macrocyclic polycarboxylate-hydrophobic ammonium carriers, *Tetrahedron Lett.,* 29, 1153, 1988.

46. **Shinkai, S., Koreishi, H., Ueda, K., Arimura, T., and Manabe, O.,** Molecular design of calixarene-based uranophiles which exhibit remarkably high stability and selectivity, *J. Am. Chem. Soc.,* 109, 6371, 1987.

47. **Matsuoka, H., Aizawa, M., and Suzuki, S.,** Uphill transport of uranium across a liquid membrane, *J. Membr. Sci.,* 7, 11, 1980; **Huang, T. C. and Huang, C. T.,** The mechanism of transport of uranyl nitrate across a solid supported liquid membrane using tributyl phosphate as mobile carrier, *J. Membr. Sci.,* 29, 295, 1986.

48. **Shimidzu, T. and Okushita, H.,** Carrier-mediated transport of $Ga^{3+}$ from $Ga^{3+}/Al^{3+}$ binary solutions and $Cu^{2+}/Zn^{2+}$ binary solutions through alkylated cupferron-impregnated membrane, *J. Membr. Sci.,* 27, 349, 1986.

49. **Wierenga, W.,** The total synthesis of ionophores, in *The Total Synthesis of Natural Products,* Vol. 4, ApSimon, J., Ed., John Wiley and Sons, New York, 1981, 263.

50. **Pedersen, C. J.,** Cyclic polyethers and their complexes with metal salts, *J. Am. Chem. Soc.,* 89, 7017, 1967.

51. **Kimura, K., Sakamoto, H., Kitazawa, S., and Shono, T.,** Novel lithium-selective ionophores bearing an easily ionizable moiety, *J. Chem. Soc., Chem. Commun.,* 1985, 669; **Sakamoto, H., Kimura, K., and Shono, T.,** Lithium separation and enrichment by proton-driven cation transport through liquid membranes of lipophilic crown nitrophenols, *Anal. Chem.,* 59, 1513, 1987.

52. **Bartsch, R. A., Czech, B. P., Kang, S. I., Stewart, L. E., Walkowiak, W., Charwicz, W. A., Heo, G. S., and Son, B.,** High lithium selectivity in competitive alkali-metal solvent extraction by lipophilic crown carboxylic acids, *J. Am. Chem. Soc.,* 107, 4997, 1985.

53. **Kobuke, Y., Hanji, K., Horiguchi, K., Asada, M., Nakayama, Y., and Furukawa, J.,** Macrocyclic ligands composed of tetra-hydrofuran for selective transport of monovalent cations through liquid membrane, *J. Am. Chem. Soc.,* 98, 7414, 1976.

54. **Lehn, J.-M. and Sauvage, J.-P.,** Cation and cavity selectivities of alkali and alkaline-earth "cryptates", *J. Chem. Soc., Chem. Commun.,* 1971, 440.

55. **Hiratani, K.,** A new synthetic ionophore exhibiting selective Li⁺ transport, *Chem. Lett.,* 1982, 1021; **Hiratani, K., Taguchi, K., Sugihara, H., and Iio, K.,** The synthesis of Li⁺-selective polyether carriers and their behavior in cation transport through liquid membranes, *Bull. Chem. Soc. Jpn.,* 57, 1976, 1984.

56. **Hiratani, K.,** Spectroscopic behaviors of the alkali salts of noncyclic polyethers: evidence for the interaction between the end groups, *Bull. Chem. Soc. Jpn.,* 58, 420, 1985.

57. **Hiratani, K., Taguchi, K., Sugihara, H., and Iio, K.,** Synthesis and properties of noncyclic polyether compounds. XV. Noncyclic polyether carriers exhibiting potassium ion-selective transport through liquid membranes: their structures and cation-selectivity, *J. Membr. Sci.,* 35, 91, 1987.

58. **Hiratani, K., Taguchi, K., Sugihara, H., and Okada, T.,** A noncyclic neutral carrier exhibiting excellently Li⁺-selective transport, *Chem. Lett.,* 1986, 197.

59. **Ueno, K., Hiratani, K., Taguchi, K., Okada, T., and Sugihara, H.,** Crystal structure of 1:1 complex of 1,3-bis(8-quinolyloxy)propane and lithium perchlorate, *Chem. Lett.,* 1987, 949.

60. **Shanzer, A., Samuel, D., and Korenstein, R.,** Lipophilic lithium ion carriers, *J. Am. Chem. Soc.,* 105, 3815, 1983.

61. **Heumann, K. G.,** Isotopic separation in systems with crown ethers and cryptands, in *Topics in Current Chemistry,* Vol. 127, Springer-Verlag, Berlin, 1985, 77.

62. **Nishizawa, K. and Watanabe, H.,** Intrinsic isotope effect of cryptand($2_B$,2,1) to Li-6 and Li-7, *J. Nucl. Sci. Technol.,* 23, 843, 1986.

63. **Symons, E. A.,** Lithium isotope separation: a review of possible techniques, *Sep. Sci. Technol.,* 20, 633, 1985.

64. **Fujie, M., Fujii, Y., Nomura, M., and Okamoto, M.,** Isotope effects in electrolytic formation of lithium amalgam, *J. Nucl. Sci. Technol.,* 23, 330, 1986.

65. **Nishizawa, K. and Hiratani, K.,** unpublished results, 1987.

66. **Tachibana, K., Koga, K., Mori, K.,** Chromatrographic preparation of optical isomers, *Kemikaru Enjiniyaringu,* 33, 380, 1988.

67. **Scheper, T., Halwachs, W., Schugerl, K.,** Production of L-amino acid by continuous enzymatic hydrolysis of DL-amino acid methyl ester by the liquid membrane technique, *Chem. Eng. J.* (Lausanne), 29, B31, 1984.

68. **Scheper, T., Werner, H., and Karl, S.,** Production of L-amino acids by continuous enzyme-catalyzed D,L-amino ester hydrolysis using liquid membrane emulsions, *Chem. Ing. Tech.,* 54, 696, 1982.

69. **Makryaleas, K., Scheper, T., Schugerl, K., and Kula, M. -R.,** Enzyme-catalyzed preparation of L-amino acids with continuous coenzyme regeneration by using liquid membrane emulsions, *Ger. Chem. Eng.,* 8, 345, 1985.

70. **Hikita, H. and Ishikawa, H.,** Separation of optical isomers by using a membrane reactor, *Kemikaru Enjiniyaringu,* 31, 292, 1986.

71. **Cussler, E. L.,** Membranes which pump, *AIChE J.,* 17, 1300, 1971.

72. **Klein, J., Baker, J. A., and Cussler, E. L.,** Optical isomer separation by diffusion, *Ind. Eng. Chem. Fundam.,* 10, 183, 1971.

73. **Behr, J.-P., and Lehn, J.-M.,** Transport of amino acids through organic liquid membranes, *J. Am. Chem. Soc.,* 95, 6108, 1973.

74. **Newcomb, M., Helgeson, R. C., and Cram, D. J.,** Enantiomer differentiation in transport through bulk liquid membranes, *J. Am. Chem. Soc.,* 96, 7367, 1974.

75. **Newcomb, M., Toner, J. L., Helgeson, R. C., and Cram, D. J.,** Host-guest complexation. 20. Chiral recognition in transport as a molecular basis for a catalytic resolving machine, *J. Am. Chem. Soc.,* 101, 4941, 1979.

76. **de Jong, F. and Reinhoudt, D. N.,** *Stability and Reactivity of Crown-Ether Complexes,* Academic Press, London, 1981, 103.

77. **Gokel, G. W. and Korzeniowski, S. H.,** *Macrocyclic Polyether Synthesis,* Springer-Verlag, Berlin, 1982.

78. **Stoddart, J. F.,** Chiral crown ethers, in *Topics in Stereochemistry,* Eliel, E. L. and Wilen, S. H., Eds., John Wiley and Sons, New York, 1987, 207.

79. **Nakazaki, M., Yamamoto, K., Ikeda, T., Kitsuki, T., and Okamoto, Y.,** Synthesis and chiral recognition of novel crown ethers incorporating helicene chiral centers, *J. Chem. Soc., Chem. Commun.,* 1983, 787.

80. **Yamamoto, K., Noda, K., Okamoto, Y.,** Synthesis and chiral recognition of optically active crown ethers incorporating a *trans* doubly-bridged ethylene framework, *J. Chem. Soc., Chem. Commun.,* 1985, 1421.

81. **Yamamoto, K., Fukushima, H., Okamoto, Y., Hatada, K., and Nakazaki, M.,** Synthesis and chiral recognition of optically active crown ethers incorporating biphenanthryl moiety as a chiral center, *J. Chem. Soc., Chem. Commun.,* 1984, 1111.

82. **Naemura, K., Fukunaga, R., and Yamanaka, M.,** Synthesis and enantiomer recognition of novel crown ethers containing the 5,6,11,12-Tetrahydro-5,11-methanodibenzo[a,e]cyclo-octene subunit as the chiral center, *J. Chem. Soc., Chem. Commun.,* 1985, 1560.

83. **Sugiura, M. and Yamaguchi, T.,** Coupled transport of amino acids through liquid membranes supported by microporous films, *Nippon Kagaku Kaishi,* 1983, 854; **Sugiura, M. and Yamaguchi, T.,** Coupled transport of amino acids through poly(vinyl chloride)-trialkyl phosphate membranes, *Nippon Kagaku Kaishi,* 1982, 1428.

84. **Yamaguchi, T., Nishimura, K., Shinbo, T., and Sugiura, M.,** Chiral crown ether-mediated transport of phenylglycine through an immobilized liquid membrane, *Maku(Membrane),* 10, 178, 1985; **Sugiura, M. and Yamaguchi, T.,** Effect of side chain in macrocyclic carriers on the carrier-mediated transport of amino esters across supported liquid membranes, *Sep. Sci. Technol.,* 19, 623, 1984.

85. **Yamaguchi, T., Nishimura, K., Shinbo, T., Kikkawa, M., Sugiura, M., Kamo, N., and Kobatake, Y.,** Supported liquid membrane for enantiomeric resolution of amino acids, presented at Int. Congr. Membranes and Membrane Processes, Tokyo, June 8 to 12, 1987, 823 (15-P15, a missing page).

86. **Yamaguchi, T., Sugiura, M., Shimakura, Y., Kamo, N., and Kobatake, Y.,** Symport of amino acids through a polymer-supported liquid membrane: evidence for diffusion controlled transport, *Koubunshi Ronbunshu,* 43, 787, 1986.

87. **Yamaguchi, T., Nishimura, K., Shinbo, T., and Sugiura, M.,** Amino acid transport through supported liquid membranes: mechanism and its application to enantiomeric resolution, *Bioelectrochem. Bioenerg.,* 20, 109, 1988.

88. **Yamaguchi, T., Nishimura, K., Shinbo, T., and Sugiura, M.,** Enantiomer resolution of amino acids by a polymer-supported liquid membrane containing a chiral crown ether, *Chem. Lett.,* 1985, 1549.

89. **CROWNPAK:** Daicel. Chem. Ind., Ltd., Osaka, Japan, 1988.

90. **Rebek, J., Jr.,** Model studies in molecular recognition, *Science,* 235, 1478, 1987; Rebek, J., Jr., Askew, B., Nemeth, D., and Parris, K., Convergent functional groups. 4. Recognition and transport of amino acids across a liquid membrane, *J. Am. Chem. Soc.,* 109, 2432, 1987.

91. **Benzing, T., Tjivikua, T., Wolfe, J., and Rebek, J., Jr.,** Recognition and transport of adenine derivatives with synthetic receptors, *Science,* 242, 266, 1988.

92. **Aoyama, Y., Yamagishi, A., Asagawa, M., Toi, H., and Ogoshi, H.,** Molecular recognition of amino acids: two-point fixation of amino acids with bifunctional metalloporphyrin receptors, *J. Am. Chem. Soc.,* 110, 4076, 1988.

93. **Shinbo, T., Nishimura, K., Yamaguchi, T., and Sugiura, M.,** Uphill transport of monosaccharides across an organic liquid membrane, *J. Chem. Soc., Chem. Commun.,* 1986, 349.

94. **Shinbo, T., Nishimura, K., Yamaguchi, T., and Sugiura, M.,** Uphill transport of saccharides and nucleosides across liquid membranes, presented at Int. Congr. Membranes and Membrane Processes, Tokyo, June 8 to 12, 1987, 813.

95. **Sverker Högberg, A. G.,** Two stereoisomeric macrocyclic resolucinol-acetaldehyde condensation products, *J. Org. Chem.,* 45, 4498, 1980.

96. **Sverker Högberg, A. G.,** Stereoselective synthesis and DNMR study of two 1,8,15,22-tetra-phenyl[l₄]metacyclophan-3,5,10,12,17,19,24,26-octols, *J. Am. Chem. Soc.,* 102, 6046, 1980.

97. **Schneider, H.-J., Gütes, D., and Schneider, U.,** A macrobicyclic polyphenoxide as receptor analogue for choline and related ammonium compounds, *Angew. Chem. Int. Ed. Engl.,* 25, 647, 1986.

98. **Aoyama, Y., Tanaka, Y., Toi, H., and Ogoshi, H.,** Polar host-guest interaction. Binding of nonionic polar compounds with a resolucinol-aldehyde cyclooligomer as a lipophilic polar host, *J. Am. Chem. Soc.,* 110, 634, 1988.

99. **Lehn, J.-M., Moradpour, A., and Behr, J.-P.,** Antiport regulation of carrier mediated chiroselective transport through a liquid membrane, *J. Am. Chem. Soc.,* 97, 2532, 1975.

100. **Diederich, F. and Dick, K.,** A new water-soluble macrocyclic host of the cyclophane type: host-guest complexation with aromatic guests in aqueous solution and acceleration of the transport of arenes through an aqueous phase, *J. Am. Chem. Soc.,* 106, 8024, 1984.

101. **Vögtle, F., Müller, W. M., Werner, U., and Losensky, H.-W.,** Selective molecular identification and separation of isomeric and partially hydrogenated arenes, *Angew. Chem. Int. Ed. Engl.,* 26, 901, 1987.

102. **Seward, E. and Diederich, F.,** Redox-dependent complexation ability of flavin-hosts in aqueous solution, *Tetrahedron Lett.,* 28, 5111, 1987.

103. **Diederich, F.,** Complexation of neutral molecules by cyclophane hosts, *Angew. Chem. Int. Ed. Engl.,* 27, 362, 1988.

104. **Harada, A. and Takahashi, S.,** Transport of neutral azobenzene derivatives by methylated cyclodextrins, *J. Chem. Soc., Chem. Commun.,* 1987, 527.

105. **Armstrong, D. W. and Jin, H. L.,** Enrichment of enantiomer and other isomers with aqueous liquid membranes containing cyclodextrin carriers, *Anal. Chem.,* 59, 2237, 1987.

106. **Dahuron, L. and Cussler, E. L.,** Protein extractions with hollow fibers, *AIChE J.,* 130, 34, 1988.

107. **Abbott, N. L. and Hatton, T. A.,** Liquid-liquid extraction for protein separations, *Chem. Eng. Prog.,* 84(8), 31, 1988.

108. **Hatton, T. A.,** Extraction of proteins and amino acids using reversed micelles, in *Ordered Media in Chemical Separation, ACS Symposium Series 342,* Hinze, W. L. and Armstrong, D. W., Eds., Am. Chem. Soc., Washington, D.C., 1987, 170.

109. **Armstrong, D. W. and Li, W.,** Highly selective protein separations with reversed micellar liquid membranes, *Anal. Chem.,* 60, 86, 1988.

*New Trends in Liquid Membrane Science*

Chapter 7.1

# EMULSION AND SUPPORTED LIQUID MEMBRANES

**Reed M. Izatt, Jerald S. Bradshaw, John D. Lamb, and Ronald L. Bruening**

## TABLE OF CONTENTS

# I. INTRODUCTION

Membrane systems allow high selectivity as well as energy and material use efficiency to be obtained in separations relative to many other systems.[1-18] In particular, carrier-mediated transport allows for highly specific molecular recognition reactions to be used on a continuous basis in performing separations.[1-5,11-18] This potential for making efficient and selective separations has led to the study of several membrane types as well as many carrier molecules.[1-5,11-18] Five specific synthetic liquid membrane types have been studied. These are the bulk, thin sheet supported, hollow fiber supported, emulsion, and two-module hollow fiber supported liquid membranes. These systems are illustrated in Figure 1. The bulk and thin sheet supported liquid membranes can be used to make screening and modeling studies of specific carrier-mediated systems, but cannot be used in large scale practical separation schemes.[5,19] These systems can be used, however, in making small scale analytical separations. An example of this is the use of a thin sheet supported liquid membrane in an ion selective electrode.[13] Hence, the systems of most practical interest are the emulsion and hollow fiber supported liquid membrane types. The advantages and disadvantages in using these two membrane types is one of the subjects of this chapter.

Perhaps the greatest need in the design of carrier-mediated liquid membrane systems is that of obtaining proper carrier molecules. These carriers must be retained in the membrane and be highly selective in performing the desired separation so that the necessary efficiency can be maintained in the membrane system. Macrocycles such as crown ethers are ideal candidates for use as carrier reagents since they have highly hydrophobic exteriors, but the inner hydrophillic cavity has high affinity for and selectivity among cations.[5,12-20] Macrocycles have also been designed with selective interactions among anions and neutral organic molecules.[21-24] The designed use of macrocycles in membranes is aided by the observation that selectivity orders in membrane systems are often similar to those found for cation binding by the same ligands in homogeneous solution.[5] The study of carrier-mediated membrane transport in our laboratory has centered on using macrocycles as carrier reagents.[5,17-19] Hence, as the properties of different membrane types and specific chemical applications are described in this chapter, examples involving macrocycles as carriers in specific systems will be used.

In this chapter, the emulsion, hollow fiber supported, and two-module hollow fiber systems will be compared. First, the advantages and disadvantages of the different systems will be discussed. Then, several specific separations will be examined in each of the systems to illustrate these advantages and disadvantages. Examination of separations in several specific systems will also illustrate interesting characteristics of both the membrane systems and the carrier molecules. Finally, the potential practical applications of these carrier-mediated liquid membranes will be presented and discussed.

# II. PRACTICAL MEMBRANE SYSTEM ADVANTAGES AND DISADVANTAGES

## A. EMULSION LIQUID MEMBRANE

This system[5,16,19] has a very thin membrane and immense surface area resulting in rapid transport. Desired species are also concentrated by a large factor upon transport from the source to the receiving phase. This concentration is due to the large aqueous source phase to receiving phase volume ratio in these systems. This large ratio results when the organic phase-receiving phase emulsion volume is added to a much larger source phase volume. In the emulsion membrane system, both the membrane solvent and carrier molecule need to be only moderately hydrophobic in order to maintain membrane stability and to retain the carrier molecule in the membrane phase. The lowered hydrophobicity requirements (relative

**(a) Bulk Liquid Membrane**

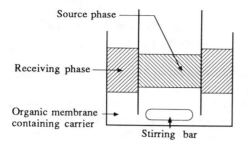

**(b) Thin Sheet Supported Liquid Membrane**

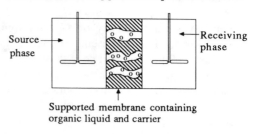

Supported membrane containing
organic liquid and carrier

**(c) Hollow Fiber Supported Liquid Membrane**

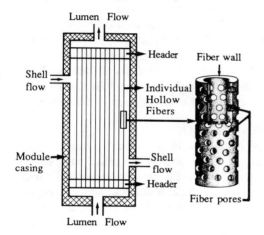

**(d) Emulsion Liquid Membrane**

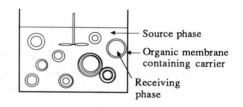

**(e) Two–Module Hollow Fiber Supported Liquid Membrane**

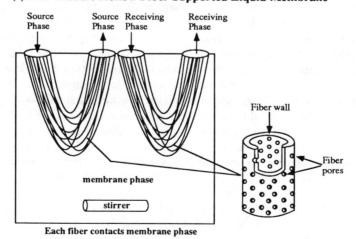

FIGURE 1.   Liquid membrane types.

to the hollow fiber system described in Section II.B.) are due to the moderate overall aqueous phase to organic phase volume ratio and the added stability of the membrane due to emulsification. Compared to the hollow fiber system, the volume ratio is not large since the volumes of the receiving phase and the organic phase are equal and since large source phase volumes cannot be used and still maintain a large ratio of membrane surface area to aqueous source phase volume. The latter condition is required to maintain the high transport rate of the emulsion system as discussed previously. Another advantage of the emulsion system is that any surface activity of a proton-ionizable carrier is relatively unimportant since an emulsion is already present. Surface activity will be seen to be a potential problem with supported liquid membrane systems.

Most of the disadvantages of the emulsion system are related to the formation of the emulsion itself. These disadvantages can be summarized as follows. First, any factors which adversely affect emulsion stability during system operation must be controlled. These factors include water phase pH and ionic strength values, as well as physical forces present in system operation. Second, after the transport and separation capacity of the system is exhausted, the emulsion must be broken down to recover the receiving phase containing the separated product. Often, emulsion stability during system operation must be balanced against ease of emulsion breakdown in the design of an emulsion liquid membrane system. Third, the emulsion must be broken down in order to replenish the carrier in the membrane, if this becomes necessary during large scale system operation. This problem can be overcome by using more hydrophobic macrocycles, but, of course, this eliminates the emulsion advantage of the system. The final disadvantage of the emulsion system is that, if the membrane does not remain intact during operation for any reason, the source and receiving phases are readily mixed together destroying any separation already concluded or in progress in the system.

## B. HOLLOW FIBER SUPPORTED LIQUID MEMBRANE

The surface area and membrane thickness of the hollow fiber system[18,19] result in the transport of desired species being rapid, but not as rapid as that of the emulsion system. Furthermore, the design of a large surface area into a hollow fiber system requires the use of a large number of long fibers. The fibers and the attendant fiber housing apparatus become a large capital cost in the use of such a system. The source and receiving phases, however, are much more easily introduced and recovered from this system. Furthermore, the entire source and receiving phases are not in contact with the membrane at a particular instant. Hence, close monitoring of membrane leakage will allow for a minimal contamination of the source and/or receiving phase if a membrane break occurs. The aqueous phases not in contact with the membrane at the time leakage begins can quickly be sealed off from the membrane system. The hollow fiber system, like the emulsion system, can also be designed to concentrate the separated species by flowing greater volumes of the aqueous source phase than the receiving phase through the system. In fact, the concentration factor is more readily controlled by a proper design in the hollow fiber system. Surface area to source phase volume ratio requirements can limit the natural concentration factor in the emulsion system.

A much smaller inventory of membrane solvent and carrier is needed in the hollow fiber system than in the emulsion system since the volume ratios of both aqueous phases to the membrane are large. However, the large volume ratios are also a disadvantage in that very hydrophobic membrane solvents and carrier reagents are required to maintain membrane integrity and operate the separation system. The hollow fiber system also has the disadvantage that the hollow fiber module must be cleaned of the aqueous phases and any contaminated or degraded membrane phase in order for either fresh carrier or membrane solvent reagents to be added to a system in use. The final disadvantage of the hollow fiber system is that the pores of the hollow fiber solid support can be fouled due to surface effects as well as particles which enter the system.

## C. TWO-MODULE HOLLOW FIBER SUPPORTED LIQUID MEMBRANE

The two-module hollow fiber system[25] was designed to allow for the minimization of aqueous to organic phase volume ratios, while maintaining a relatively high transport rate. This design makes possible the use of membrane solvents and carrier reagents of low hydrophobicities. Furthermore, additional membrane solvent and carrier reagent are easily added to replace those lost during operation. This ease of replacement is a large advantage over either the regular hollow fiber or the emulsion membrane system. Disadvantages of using this system are (1) the bulk of the large membrane phase must be effectively stirred to maximize transport rates, and (2) any additional boundary layers and the required transport through two sets of hollow fiber modules provide for slower transport rates compared to the other two systems. The other advantages and disadvantages of this system are similar to those of the regular hollow fiber membrane system, i.e., the desired aqueous volume ratio is readily controlled, both aqueous phases are easily introduced and recovered from the system, leakage emergencies do not contaminate an entire product, fouling can be a problem, and a large capital cost can be involved.

# III. SEPARATION SYSTEM EXAMPLES

The advantages and disadvantages of the three potentially practical emulsion, hollow fiber supported, and two-module hollow fiber supported liquid membrane systems are best described by examining data from some specific separations. The examples presented will be confined to cation separations using macrocyclic carriers. However, the principles illustrated also apply to the transport of other species either by macrocycles or by other types of carriers. Cation and co-anion transport via a neutral carrier as well as cation transport with counter transport of protons and a proton-ionizable carrier will be examined. The examples presented also illustrate important general characteristics of liquid membrane systems.

## A. SEPARATION OF ALKALI CATIONS WITH CROWN ETHERS AS CARRIERS

Members of the crown ether class of macrocyclic compounds interact selectively with particular alkali metal cations primarily according to the fit of the cation into the crown cavity.[5,20] This type of selectivity has been observed in homogeneous solution,[20] solvent extraction,[26,27] and liquid membrane[5] systems. Macrocycles containing an 18-crown-6 macrocyclic core (Figure 2) show size-based selectivity for $K^+$ over the other alkali metal cations. Crown ethers having other core sizes show interaction selectivity for other alkali cations.[20] The system conditions necessary to perform 18-crown-6 mediated separations of $K^+$ are illustrative of the factors that must be considered in the design of emulsion and hollow fiber membranes. Addition of hydrophobic substituent groups to the 18-crown-6 core is a necessary part of macrocycle design, Figure 2a (where R = 1-hydroxyheptyl), when the emulsion and hollow fiber liquid membrane types are used in order to maintain the macrocycle in the membrane.[5,18,19,28] This is because of the large aqueous to organic phase volume ratios for these systems.

The large aqueous to membrane phase volume ratios of these systems also require the use of a fairly hydrophobic membrane solvent in order to maintain membrane integrity.[5,18,19,28] At very high aqueous to organic volume ratios, extremely hydrophobic solvents are needed to maintain membrane integrity. For example, membrane stability as a function of time was examined with a thin sheet supported liquid membrane system where the aqueous to organic volume ratio was 1333:1.[17] In Table 1, the duration of membrane stability with this particular system is shown as a function of solvent boiling point and, most importantly, aqueous solubility. The extended alkyl chain in comparison to toluene gives phenylhexane its favorable properties. It is obvious that when the aqueous to organic volume ratio is very large, a solvent of high hydrophobicity such as phenylhexane is required for membrane stability. The less hydrophobic solvents toluene and 1-octanol maintain stable membranes in emulsion and hollow fiber systems, which have smaller aqueous to organic volume ratios of 11:1 and 80:1, respectively.[5,18]

Effective partitioning of the macrocycle to the organic solvent over water as organic

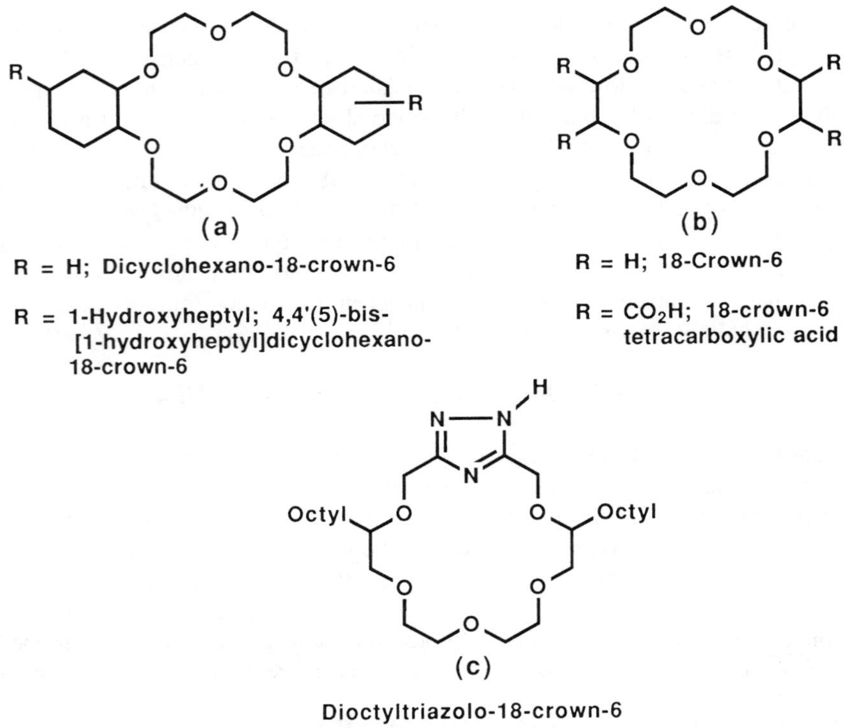

(a)

R = H; Dicyclohexano-18-crown-6

R = 1-Hydroxyheptyl; 4,4'(5)-bis-
[1-hydroxyheptyl]dicyclohexano-
18-crown-6

(b)

R = H; 18-Crown-6

R = $CO_2H$; 18-crown-6
tetracarboxylic acid

(c)

Dioctyltriazolo-18-crown-6

FIGURE 2.  Macrocycles containing an 18-crown-6 core.

## TABLE 1
### The Effect of Solvent Type on Thin Sheet Supported Liquid Membrane[a] Stability

| Membrane solvent | Boiling point[b] (C°) | Water solubility[c] (wt. %) | Membrane stability[a] |
|---|---|---|---|
| CHCl₃ | 61.7 | 0.71 | Minutes |
| Toluene | 110.6 | 0.063 | Hours |
| Dichlorobenzene | 180.6 | 0.0145 | Hours |
| Phenylhexane | 227 | d | Stable |

[a]  A 200 ml 0.1 *M* KNO₃, NaNO₃, LiNO₃/0.3 ml solvent on
Celgard 2400 polypropylene/200 ml H₂O membrane.[17]
Volume ratio of aqueous phase to organic phase is 1333:1.
[b]  Ref. 29.
[c]  Ref. 30.
[d]  No value reported.

solvents of increasing hydrophobicity are used requires macrocycles of increasing hydrophobicity as well. This point is illustrated in Table 2 by the partition coefficient, $\underline{K}_p$, data for dicyclohexano-18-crown-6 as a function of organic membrane solvent.[28,31,32] In Table 2, the $\underline{K}_p$ values are seen to decrease as solvent hydrophobicity increases. This decrease indicates that the affinity of the moderately hydrophobic dicyclohexano-18-crown-6 for water increases as solvent hydrophobicity increases. Thus, maintenance of membrane integrity in systems having large aqueous:organic volume ratios requires both a very hydrophobic solvent and a very hydrophobic macrocycle. Macrocycle hydrophobicity is increased by adding

## TABLE 2
### Partition Coefficient ($\underline{K}_p$)[a] of Dicyclohexano-18-crown-6[b] as a Function of Increasingly Hydrophobic Organic Solvent

| Solvent | $\underline{K}_p$[a] |
|---|---|
| $CH_2Cl_2$ | 713 |
| $CHCl_3$ | 454 |
| $C_2H_4Cl_2$ | 196 |
| $CCl_4$ | 108 |
| $C_2H_4Cl_4$ | 90 |
| Toluene | 13 |
| Phenylhexane | c |

[a] Constant for Ligand$_{aq}$ = Ligand$_{org}$ partitioning.[31,32]
[b] Mixture of cis-syn-cis and cis-anti-cis isomers.
[c] Value is too small for measurement.

appropriate substituent groups. However, the selection of the hydrophobic group(s) is important. The addition of alkyl and cycloalkyl substituent groups to macrocycles has been found to increase their hydrophobicity with minimal reduction of their complexing ability.[20] On the other hand, benzo, vinyl, and other electron withdrawing hydrophobic substituent groups add necessary hydrophobicity, but also reduce the macrocycle complexing power.[5,20] Hence, alkyl and cycloalkyl substitution is preferred.

An example of appropriate hydrophobic substituent substitution was the use of 4,4′(5)-bis[1-hydroxyheptyl]dicyclohexano-18-crown-6 (Figure 2a) in the thin sheet supported liquid membrane containing phenylhexane as membrane solvent and with an aqueous:organic volume ratio of 1333:1 as described above.[17] This macrocycle has a $\underline{K}_p$ value of 15,000 in favor of phenylhexane over water. Hence, the macrocycle is maintained in the membrane. In a membrane system consisting of a 200 ml 0.1 $M$ LiNO$_3$, 0.1 $M$ NaNO$_3$, and 0.1 $M$ KNO$_3$ source phase; a 0.3 ml of 0.5 $M$ macrocycle in phenylhexane on Celgard 2400 polypropylene membrane phase; and a 200 ml water receiving phase, the KNO$_3$, NaNO$_3$, and LiNO$_3$ flux values were 3.50, 0.59, and 0 × 10$^{-8}$ mol·s$^{-1}$·m$^{-2}$, respectively. Thus, with a stable membrane and a macrocycle which is maintained in the membrane phase, the desired and expected high selectivity for K$^+$ over other alkali metal ions was obtained using an 18-crown-6-type macrocycle.

Membrane stability and macrocycle loss from the membrane are not the only factors to be considered in choosing a membrane solvent for making a particular separation. While macrocycle selectivity is almost exclusively a function of cavity type and the addition of alkyl and cycloalkyl groups does not reduce macrocycle binding ability, membrane solvent type greatly affects the ease of extraction of the cation and its accompanying anion in the case of a neutral carrier. This principle can be illustrated using the data in Table 3.[28] While the difference in the log $\underline{K}_{ex}$ values for pairs of cations which is indicative of selectivity shows only minor variations, the drop in equilibrium constant ($\underline{K}$) for cation-macrocycle interaction as a function of solvent of over 3 log $\underline{K}$ units is significant. For a given cation, the $\underline{K}_{ex}$ values decrease as the organic solvent becomes more hydrophobic. The decrease is not gradual, but drops off significantly at a certain point. Hence, in designing a membrane system usable with a large aqueous to organic solvent volume ratio the problem of a large drop in log $\underline{K}_{ex}$ values must be considered.

## B. USE OF RECEIVING PHASE COMPLEXING AGENTS

It is usually desirable to concentrate transported species in liquid membrane systems at the same time that a separation is being performed. In the practical emulsion and hollow

## TABLE 3
### Extraction Equilibrium Constants ($\underline{K}_{ex}$)as a Function of Solvent Type with Dicyclohexano-18-crown-6[a] as Macrocycle

| Solvent | Log $\underline{K}_{ex}$[b] | | |
|---|---|---|---|
| | $Pb^{2+}$ | $Sr^{2+}$ | $K^+$ |
| $CH_2Cl_2$ | 5.4 | 3.3 | 2.3 |
| $C_2H_4Cl_2$ | 5.3 | 3.2 | 2.3 |
| $CHCl_3$ | 5.1 | 3.2 | 1.9 |
| $C_2H_2Cl_4$ | 4.8 | 3.3 | 2.6 |
| Toluene | 3.9 | 0.9 | 0.2 |
| $CCl_4$ | 2.1 | <0.1 | <0.1 |

[a]  Mixture of cis-syn-cis and cis-anti-cis isomers.
[b]  Constant for macrocycle$_{(org)}$ + cation $(NO_3)_{n(aq)}$ = complex$_{(org)}$ interaction.[28]

## TABLE 4
### KNO₃ *vs.* NaNO₃ Transport in an Emulsion Liquid Membrane[a] as a Function of Receiving Phase Complexing Agent

| Complexing agent | Time elapsed (min) | Percent transport(%) | |
|---|---|---|---|
| | | $K^+$ | $Na^+$ |
| None | 30 | 9 | 0 |
| $Li_4P_2O_7$[b] | 30 | 11 | 3 |
| 18-crown-6[b] | 10 | 0 | 0 |
| $(CO_2H)_4$ 18-crown-6[c] | 10 | 17 | 1 |
| $(CO_2H)_4$ 18-crown-6[c] | 30 | 95 | 34 |

[a]  A 0.001 *M* KNO₃ and NaNO₃/0.02 *M* dicyclohexano-18-crown-6 in toluene, 3% v/v sorbitan monooleate/receiving phase membrane.[33]
[b]  0.05 *M*.
[c]  0.03 *M*.

fiber liquid membrane systems, the use of a receiving phase of smaller volume than the source phase is ideal for this purpose. When neutral macrocycles are used to transport a cation selectively, an anion must accompany the resulting macrocycle-cation complex. In the absence of a reverse pH gradient or a cation complexing agent in the receiving phase, transport will cease when equilibrium concentrations of the transported salt are present in both the source and receiving phases. An example of this principle is seen in the first line of Table 4 for an emulsion membrane system using dicyclohexano-18-crown-6 (Figure 2a) as carrier and with a source to receiving phase volume ratio of 10:1. After 30 min, equilibrium is reached with 9% of the KNO₃ transported. This transported amount is in good agreement with that expected (10%) from the volume ratio. Cessation of KNO₃ transport occurs because the final concentration of KNO₃ in both the source and receiving phases is now equal due to the aqueous volume ratio of the system, and thus, there is no longer a concentration gradient in KNO₃ across the membrane. Selective transport of KNO₃ over NaNO₃ using dicyclohexano-18-crown-6 as a carrier occurs when an initially pure water phase is used as a receiving phase (first line of Table 4). No transport of $NaNO_3$ is found up to 30 min because of the lack of interaction between $Na^+$ and the 18-crown-6-sized macrocycle.[33]

The total amount of $KNO_3$ transported at a given time can be increased in the emulsion system by incorporating a complexing agent for $K^+$ into the receiving phase to decrease the $K^+$ concentration in that phase. This procedure allows for an overall concentration gradient in free $KNO_3$ to be maintained across the membrane, and transport continues until this gradient disappears. The effect on transport of different receiving phase complexing agents is illustrated using the data in Table 4. Pyrophosphate ion is an excellent complexing agent for many cations, but not for the alkali metal cations. For example, the log $K$ value for $K^+$-$P_2O_7^{4-}$ interaction in aqueous solution ranges from only 0.8 to 2.1 at various ionic strengths.[34] This low complexing ability is the reason for the very slight $KNO_3$ equilibrium transport enhancement observed when $P_2O_7^{4-}$ is used in the receiving phase of this particular emulsion system. A similar explanation can be given for the low $Na^+$ transport. There are few reagents which complex alkali cations to any significant degree in aqueous solution. The macrocycle 18-crown-6 (Figure 2b) does complex with the alkali metal cations and could be used in the receiving phase to complex the transported cations. However, as shown in Table 4, 18-crown-6 in the receiving phase does not increase transport of $K^+$. This is because 18-crown-6 is also soluble in the membrane phase and quickly equilibrates through the organic membrane between the two aqueous phases. A solution to this dilemma is to increase the hydrophilicity of the receiving phase reagent. The carboxylic acid derivative of 18-crown-6 (Figure 2b) is easily deprotonated. The resulting anionic species is extremely hydrophilic, is retained in the receiving phase, and strongly complexes $K^+$. Hence, the transport rate and equilibrium amount of transported $K^+$ are greatly enhanced. However, the time at which the emulsion membrane should be broken down and the $K^+$ recovered must be monitored since, when almost all of the $K^+$ has been transported, the $Na^+$ transport rate begins to increase and selectivity becomes progressively poorer (see Table 4, line 5).[33]

Other receiving phase complexing agents can be used when various cations or groups of cations are being transported selectively in membrane systems. The use of $S_2O_3^{2-}$ for Ag(I), Cd(II), Hg(II), and Pb(II) transport and $P_2O_7^{4-}$ or $EDTA^{4-}$ for the enhancement of transport of a number of cations are examples.[5,34] These anions are highly charged, extremely hydrophilic, and hence, are maintained in the receiving phase of a membrane system. The choice of receiving phase reagent must be carefully considered, however. A reagent with interaction selectivity for the same cation(s) as is the case for the carrier molecule can be used to enhance membrane system selectivity as well as transport rate and total transport efficiency. On the other hand, variation in interaction selectivity order between a receiving phase complexing agent and membrane carrier may completely destroy a desired selectivity.

## C. CO-ANION EFFECTS ON MEMBRANE TRANSPORT USING A NEUTRAL CARRIER MOLECULE

When neutral macrocycle carriers are used to transport cations, anions must accompany the cation-carrier complex in order to maintain electrical neutrality. Hence, the extraction of the anions can be important in determining transport rates, selectivities, and the total efficiency of transport. The effect of anions on transport is twofold. First, the less hydrophilic the anion the greater the rate of transport. This factor is independent of the cation. Thus, anion hydrophilicity affects cation transport rates similarly, but does not affect selectivity or the overall efficiency of transport. The variation of transport rate with anion hydrophobicity is large. In a study of $K^+$ transport by 18-crown-6 as a function of anion type in a bulk liquid membrane system,[35] transport rates of the fastest (picrate ion) and slowest ($F^-$) systems differed by nearly $10^8$. The use of $K^+$ for this illustration is purposeful in that ion pairing of $K^+$ with most anions is negligible and, thus, other effects of anion type on $K^+$ transport are unimportant.[5] Second, interactions between cations and anions can affect membrane

## TABLE 5
### Use of Co-Anion Type and Concentration to Separate Zn(II), Cd(II), and Hg(II) Using 18-Crown-6 Derivatives in Emulsion and Hollow Fiber Supported Liquid Membranes

| Anion/[Anion][a] | Cation 1/$\alpha_2$[b] | Cation 2$\alpha_2$[b] | Selectivity Cation 1/Cation 2 Emulsion[c] | Hollow Fiber[d] |
|---|---|---|---|---|
| SCN$^-$/0.4 | Cd/0.47 | Hg/9 $\times$ 10$^{-5}$ | 75 | $\infty$[e] |
| SCN$^-$/0.004 | Cd/2 $\times$ 10$^{-4}$ | Hg/0.48 | 0[e] | 0.026 |
| SCN$^-$/0.4 | Cd/0.47 | Zn/0.26 | 1.8 | 2.0 |
| Br$^-$/0.3 | Cd/0.3 | Zn/f | $\infty$[e] | $\infty$[e] |

[a]    Anion concentration in mol/$\ell$ in the initial source phase of the membrane system.

[b]    The fraction of the total amount of source phase cation present as a neutral ion pair. Calculated using the following log $\beta_n$ values for $Cd^{2+} - nA^-$ interaction. $Cd^{2+} - SCN^-$: $\beta_1 = 1.32$, $\beta_2 = 1.99$, $\beta_3 = 2.0$, $\beta_4 = 1.9$; $Hg^{2+} - SCN^-$: $\beta_1 = 9.08$, $\beta_2 = 16.86$, $\beta_3 = 19.7$, $\beta_4 = 21.7$; $Zn^{2+} - SCN^-$: $\beta_1 = 0.71$, $\beta_2 = -1.04$, $\beta_3 = 1.2$, $\beta_4 = 1.5$; $Cd^{2+} - Br^-$: $\beta_1 = 1.57$, $\beta_2 = 2.1$, $\beta_3 = 2.6$, $\beta_4 = 2.6$; $Zn^{2+} - nBr^-$: $\beta_2$ value is too small for measurement.

[c]    Transport selectivity in a 0.001 $M$ in both cations (9 ml)/0.2 $M$ dicyclohexano-18-crown-6 in toluene (0.9 ml)/H$_2$O (0.9 ml) emulsion liquid membrane.[36]

[d]    Transport selectivity in a 0.5 $M$ in both cations (200 ml)/0.5 $M$ *bis*(1-hydroxyheptyl)dicyclohexano-18-crown-6 in phenylhexane on Celgard polypropylene (5 ml)/H$_2$O (220 ml) hollow fiber supported liquid membrane.[18,26]

[e]    Transport of the nonselective cation was undetectable.

[f]    Cation-anion interaction is not detectable.

transport rates, selectivity, and overall transport efficiency. The effect of these interactions lies in the fact that transport in these systems is driven by a concentration gradient across the membrane in the transported species. This species is the neutral cation-anion ion pair. Unlike the case of the alkali cations with the majority of anions, the extent of interactions between other anions and cations can be appreciable and can vary dramatically. In these latter cases, source phase anion types and concentrations can be controlled to effect important separations.

Examples of such separations and the selectivities obtained for Zn(II), Cd(II), and Hg(II) are given in Table 5 using emulsion and hollow fiber supported liquid membrane data. Analogs of 18-crown-6 (Figure 2a) containing sufficient hydrophobic bulk for the particular membrane system were used in these experiments. The selectivity order for cation interaction with these macrocycles, Hg(II) > Cd(II) $\approx$ Zn(II), in both solvent extraction and homogeneous solvent is identical. If the macrocyclic carrier were the only factor in determining selectivity, this would always be the selectivity order. However, in the first line of Table 5, highly selective transport of Cd(II) over Hg(II) is seen. In this experiment, 0.4 $M$ SCN$^-$ is present in the source phase so that Hg(SCN)$_4$$^{2-}$ and Cd(SCN)$_2$ are the primary species present. The neutral Cd(SCN)$_2$ species is transported preferentially over the ionic, and hence more hydrophilic, Hg(SCN)$_4$$^{2-}$. The transport efficiency, i.e., the fraction of total initial amount of Cd(II) present in the source phase which is transported, was 75 and 85% for the emulsion and hollow fiber supported liquid membranes, respectively. Equal total Cd(II) concentrations in both aqueous phases would allow for only 10 and 50% total transport since an H$_2$O receiving phase was used. The enhanced transport efficiency is due to the fact that free Cd$^{2+}$ and Cd(SCN)$^+$ are the primary species present in the receiving phase where SCN$^-$ is not in excess. Hence, the concentration gradient in Cd(SCN)$_2$ can be maintained until a large fraction of the original Cd(II) present is transported.[18,26,36]

The expected macrocycle selectivity order, Hg(II) > Cd(II), is obtained with 0.004 $M$

SCN$^-$ (line 2 of Table 5) where Cd$^{2+}$ and Hg(SCN)$_2$ are the primary species in the source phase. In this case, a Hg(II) transport efficiency of 80% was observed for the emulsion system. In this system, Hg(SCN)$_2$ is also the primary species present in the initial H$_2$O receiving phase upon transport. In fact, some precipitation of Hg(SCN)$_2$ in the receiving phase was observed and this enhanced the amount of transport possible. Fouling of the hollow fiber membrane was observed and transport could not be fully quantitated. A slight amount of excess SCN$^-$ is necessary to maintain Hg(SCN)$_2$ solubility at reasonable levels.[18,26,36]

The 2:1 Cd(II) over Zn(II) selectivity observed in the 0.4 $M$ SCN$^-$ source phase systems (line 3, Table 5) is due solely to the slight difference in affinity of the cations for SCN$^-$, since there is no macrocycle selectivity in this case. The large Cd(II) over Zn(II) selectivity when Br$^-$ is the co-transporting anion (last line, Table 5) is an example of how proper choice of source phase anion type and concentration can be used to make nearly quantitative separations using nonselective neutral carrier molecules. In this case, the Br$^-$ concentration was chosen to maximize the concentration of CdBr$_2$ while Zn-Br$^-$ interaction is negligible. Thus, high Cd(II) selectivity is found. The presence of free Cd$^{2+}$ primarily in the receiving phase upon transport results in high total transport efficiencies. The one disadvantage of using Br$^-$ rather than SCN$^-$ as the co-transporting anion is that Br$^-$ is more strongly hydrated, and hence, the transport rate of CdBr$_2$ is slower than that of Cd(SCN)$_2$ for the same CdA$_2$ concentration gradient.[18,26,36]

The similar selectivities for Cd(II), Zn(II), and Hg(II) using the different membrane types provide further evidence of how macrocyclic carriers containing similar cores, but different alkyl and cycloalkyl hydrophobic substituents, show similar selectivities. Factors other than those described above have been used with these Cd, Hg, and Zn systems. For example, S$_2$O$_3^{2-}$ as a receiving phase reagent interacts strongly and quite selectively with Hg(II) while the substitution of N or S donor atoms for O in the crown ethers enhances Hg(II) selectivity over Cd(II) and Zn(II). Combination of these factors with a slight excess concentration of SCN$^-$ in the source phase allows for virtually infinite selectivity for Hg(II) over Cd(II), Zn(II), and many other cations as well as nearly 100% transport efficiency and an extremely rapid Hg(II) transport rate.[18,26,36]

## D. USE OF PROTON-IONIZABLE MACROCYCLIC CARRIERS

Use of a proton-ionizable macrocyclic carrier in a membrane system allows for counter transport of protons rather than co-transport of anions to be involved in the cation transport mechanism. This makes it possible to eliminate the effect of anion solvation and to minimize other anion effects on the membrane transport process as described for the use of neutral macrocyclic carriers. In this way, membrane selectivity for a particular cation is controlled almost entirely by the choice of macrocyclic carrier. The main potential problem with the use of a proton-ionizable carrier is that the pH of both aqueous phases of a membrane system must be controlled. Receiving phase pH is readily controlled since the composition of this phase is synthetically designed. However, ease of source phase pH control depends largely on the matrix of the solution from which a separation is intended to be made. Any cation-anion interactions in a given source phase must be considered to be certain that desired selectivity is obtained. This is because transport is driven by a concentration gradient in the free cation being transported as well as by a reverse pH concentration gradient. An obvious advantage of using proton-ionizable carriers is that this reverse pH gradient can be used to greatly enhance overall cation transport efficiency and the rate of transport.

An example of using a particular proton-ionizable macrocylic carrier to transport selectively a single cation in a liquid membrane system is the use of a triazolo-18-crown-6 macrocycle (Figure 2c) for the selective transport of Ag(I). The dioctyltriazolo-18-crown-6 macrocycle shown in Figure 3 has sufficient hydrophobic bulk originating in the alkyl substituent groups to maintain the macrocycle in an emulsion membrane with a 10:1:1

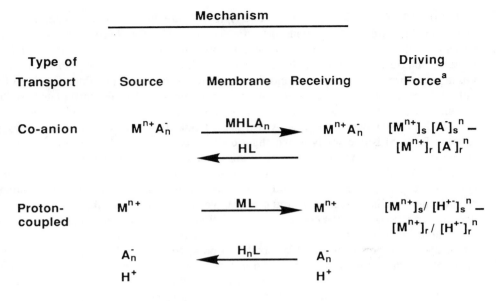

FIGURE 3.    Co-Anion (A$^-$) and proton-coupled ligand (L)-mediated transport mechanisms and associated driving forces for an M$^{n+}$ cation.

source:membrane:receiving phase volume ratio. Membrane transport of alkali and alkaline earth metal cations as well as Ag(I), Pb(II), Tl(I), Cd(II), Zn(II), Ni(II), and Fe(III) present as NO$_3^-$ salts has been examined using this triazolo macrocycle. Membrane transport of Cd(II), Zn(II), Ni(II), Fe(III), Tl(I), and the alkaline earth cations is negligible using this carrier. Alkali cation transport occurs solely by a proton-ionizable mechanism and only when sufficient base is added to the source phase to raise the pH well above 12. Because of the lack of affinity of the alkali cations for nitrogen-containing ligands, transport occurs only when the carrier is deprotonated and charge-charge interactions can occur. The pKa value for the triazole group in such macrocycles is 8.5[16] so that a high pH in the source phase is needed for alkali metal transport. The remaining cations, Ag(I) and Pb(II), are not sufficiently soluble as the nitrate salts at basic pH to be studied.

Transport of Ag(I) and Pb(II) by the triazolo crown in the emulsion liquid membrane system is observed with source phases at neutral and slightly acidic pH values. Transport of Ag(I) occurs with either a H$_2$O or HNO$_3$ receiving phase, while Pb(II) transport is found only with a H$_2$O receiving phase. Furthermore, the Pb(II) transport occurs by a co-anion transport mechanism involving NO$_3^-$. The transport of Ag(I), on the other hand, was found to occur by a proton-ionizable mechanism with an acidic receiving phase, but a NO$_3^-$ co-anion transport mechanism predominated when an H$_2$O receiving phase was used. The Pb(NO$_3$)$_2$ co-transport mechanism occurs because the triazolo carrier molecule contains only one proton-ionizable site, and macrocycles, generally, do not form 1:2 cation to macrocycle complexes, although a few examples have been cited.[5] Hence, the co-anion transport mechanism is preferred. Increasing the NO$_3^-$ concentration in the receiving phase by using 0.031 $M$ HNO$_3$ compared to that in the 0.001 $M$ Pb(NO$_3$)$_2$ source phase reduces the Pb(NO$_3$)$_2$ concentration gradient across the membrane to the point that measurable Pb(II) transport does not occur. In a comparable experiment using Ag(I), transport by a proton-ionizable mechanism does occur since an appreciable reverse phase pH gradient exists with the HNO$_3$ receiving phase. Selective transport in this case occurs because of the 1:1 matching of the cation charge in Ag(I) and the anion charge of the proton-ionizable macrocycle.[16]

This principle of matching cation charge to the number of proton-ionizable sites in the macrocyclic carrier together with judicious use of the other macrocycle selectivity design principles makes possible extremely high selectivity for $Ag^+$ over all other cations tested using the triazolo macrocycle in the emulsion membrane system containing an acidic receiving phase.[16] The high $Ag^+$ transport rate is a result of three favorable factors. First, $Ag^+$ is univalent providing a 1:1 charge match with the ionized macrocycle. Second, there is appreciable affinity of $Ag^+$ for 18-crown-6-sized macrocycles as a result of the fairly good match between the diameters of $Ag^+$ and the 18-crown-6 cavity. Finally, $Ag^+$ has high affinity for nitrogen-containing functional groups. Other cations tested have greatly reduced transport with the triazolo carrier due to the lack of one or more of these properties. The alkali metals, alkaline earth metals, and $Tl^+$ have low affinity for the triazole moiety. All of the alkali and alkaline earth cations except $K^+$, $Sr^{2+}$, and $Ba^{2+}$ along with $Fe^{3+}$, $Cd^{2+}$, $Zn^{2+}$, and $Ni^{2+}$ have low affinity for the macrocycle because they do not match the 18-crown-6 cavity size well and/or they have low affinity for nitrogen donor atoms. Finally, the bivalent cations as well as $Fe^{3+}$ have multiple charges which do not match the single charge of the ionized triazole carrier. This example shows how several features of a macrocycle can be combined in the design of a carrier for highly selective transport of a particular cation in a membrane system.

The last observed phenomenon with this particular system which requires explanation is that $Ag^+$ transport by the triazolo macrocyclic carrier occurs by a proton-ionizable mechanism with the $HNO_3$ receiving phase, but by a co-anion transport mechanism with the $H_2O$ receiving phase. Solvent extraction measurements show that both of these mechanisms are operative in the $Ag^+$-triazolo system. The log $\underline{K}_{ex}$ values using phenylhexane solvent and dioctyltriazolo-18-crown-6 (HL) are 0.59 (dimensionless) and 6.0 ($mol^{-2} \cdot \ell^2$) for the proton-ionizable and co-anion transport mechanisms, respectively.[16] The relevant reactions are given by Equations 1 and 2, respectively.

$$Ag^+_{aq} + HL_{org} = AgL_{org} + H^+_{aq} \qquad (1)$$

$$AgNO_{3aq} + HL_{org} = AgHLNO_{3org} \qquad (2)$$

Which mechanism predominates in a particular liquid membrane system depends on the respective thermodynamic driving forces associated with the two mechanisms. These driving forces are largely influenced by the composition of the membrane receiving phase. The driving force for carrier-mediated membrane transport is the difference between the concentrations of any transporting complexes at the membrane interfaces with the source and receiving phases as described in Figure 3. Interfacial equilibria determine the magnitude of the concentration gradients even before full system equilibrium is reached in membrane phase diffusion-limited transport. Full system equilibrium is reached when the concentration gradient terms ($\Delta C$) equal zero regardless of the transport limiting step and mechanism. The relative magnitudes of these concentration gradients for interfacial equilibria can be calculated from the $\underline{K}_{ex}$ expressions if both the $\underline{K}_{ex}$ values and the aqueous species concentration gradients are known. The concentration of the macrocycle in the membrane does not need to be known in order to compare the $Ag^+$ transport mechanisms since this concentration affects 1:1 stoichiometry expressions in an identical manner. Hence, the concentration gradient (driving force) term for the co-anion transport mechanism involves the product $[Ag^+]$ $\cdot [NO_3^-]$ ($mol^2 \ell^{-2}$), while that for the proton-ionizable mechanism involves the quotient $[Ag^+]/[H^+]$ (dimensionless).

In Table 6, the driving forces ($\Delta C$ or $\underline{K}_{ex}\Delta C$) are given for $Ag^+$ transport via proton-ionizable and co-anion transport mechanisms with the triazolo macrocycle as carrier in a 0.001 $M$ $AgNO_3$ in $H_2O$ source phase/phenylhexane membrane/$H_2O$ or pH 1.5 $HNO_3$ re-

## TABLE 6
### Transport Driving Forces for $Ag^+$ Transport in Emulsion Liquid Membrane Systems[a] as a Function of Receiving Phase Composition

| Receiving phase | Percent transport[c] (%) | Transport driving force[b] | | | |
|---|---|---|---|---|---|
| | | Co-anion mechanism | | Proton-ionizable mechanism[b] | |
| | | $\Delta C$[d] | $K_{ex}\Delta C$[d] | $\Delta C$[d] | $K_{ex}\beta\ \Delta cxd$ |
| $H_2O$ | 0 | $1.0 \times 10^{-6}$ | 1 | $1 \times 10^4$ | $3.0 \times 10^4$ |
| | 1 | $9.7 \times 10^{-7}$ | 0.97 | 0 | 0 |
| | 5 | $6.5 \times 10^{-7}$ | 0.65 | 0 | 0 |
| | 9 | $2.0 \times 10^{-8}$ | 0.02 | 0 | 0 |
| $HNO_3$ | 0 | $1.0 \times 10^{-6}$ | 1 | $1 \times 10^4$ | $3.9 \times 10^4$ |
| | 1 | 0 | 0 | 99 | 385 |
| | 10 | 0 | 0 | 9 | 35 |
| | 50 | 0 | 0 | 0.8 | 3.1 |
| | 70 | 0 | 0 | 0.03 | 0.1 |

a   Initially a 0.001 $M$ $AgNO_3$/dioctyltriazolo-18-crown-6 in phenylhexane/$H_2O$ or 0.031 $M$ $HNO_3$ (pH = 1.5) liquid membrane with a 10:1:1 source:membrane:receiving phase volume ratio.[16]

b   Co-anion mechanism: $Ag^+_{aq} + NO_3^-_{aq} + HL_{org} = AgHLNO_{3org}$. Proton-ionizable mechanism: $Ag^+_{aq} + HL_{org} = AgL_{org} + H^+_{aq}$.

c   Percent transport which has already occurred.

d   $\Delta C$ equals the difference in the quantities $[Ag^+] \cdot [NO_3^-]$ ($mol^2 \cdot \ell^{-2}$) and $[Ag^+]/[H^+]$ (dimensionless) between the source and receiving phases for the co-anion and proton-ionizable mechanisms, respectively. The $\underline{K}_{ex}$ values are for the reactions described in footnote b in units of $mol^{-2} \cdot \ell^2$ and dimensionless for the co-anion and proton-ionizable mechanisms, respectively. The $\underline{K}_{ex}\Delta C$ values are dimensionless in both cases.

ceiving phase emulsion liquid membrane with a 10:1:1 volume ratio for the source:membrane:receiving phase. The $\underline{K}_{ex}\Delta C$ values are also presented since they are dimensionless in all cases and can be used to compare two different cations or mechanisms directly. Transport of less than 1% is possible for both the proton-ionizable mechanism with a $H_2O$ receiving phase and the co-anion transport mechanism with a $HNO_3$ receiving phase. In fact, full system equilibrium ($\Delta C = 0$) is calculated to be reached after only 0.05 and 0.3% $Ag^+$ transport for the above situations. On the other hand, $Ag^+$ transport greater than 9 and 70% are possible for the co-anion transport mechanism with a $H_2O$ receiving phase and the proton-ionizable mechanism with a $HNO_3$ receiving phase, respectively. These calculations show how the available concentration gradient is the determining factor in the transport mechanisms observed for $Ag^+$ transport with a $H_2O$ *vs.* $HNO_3$ receiving phase. Not only do the calculations in Table 6 elucidate transport mechanisms, but the calculated and observed amounts of transport at complete system equilibrium correlate (10 *vs.* 12% for a $H_2O$ and 70 *vs.* 65% for a $HNO_3$ receiving phase) within a few percent.[16]

One final factor in using proton-ionizable macrocyclic carriers is that membrane solvents which interact even weakly with protons should be avoided. In some cases, solvents such as 1-octanol allow proton transport between the aqueous phases, although the solvent itself remains impermeable to other cations without a carrier being present.[25] It is obvious from the preceeding discussion that many factors must be considered in the design of an emulsion or hollow fiber supported liquid membrane system to perform a specific separation. However, the proper use of these parameters also allows for the design of highly efficient separations.

# IV. POTENTIAL APPLICATIONS

## A. UNUSUAL CAPABILITIES IMPARTED BY MACROCYCLE CARRIERS

Before discussing the applications possessed by emulsion and hollow fiber supported liquid membranes, it is useful to examine the peculiar capabilities which molecular recognition-based carrier molecules impart to these membranes. One area of the cation separations field in which almost all present systems have difficulty is that of separations from acidic matrices. Neutral macrocycles provide the means for the selective transport of cations in liquid membranes which contain acidic source phase matrices. Since concentration gradients are an important part of the transport driving force in such systems, high concentrations of the acid anion can enhance transport.

Three other areas in which difficulty in making separations with present methods is a problem are (1) the separation of very small amounts of a particular species when other similar components are present in much higher concentration, (2) removal of a particular species from a matrix in which the species has strong affinity for other species present in the matrix, and (3) selectively removing species with the weakest chemical or physical properties typically used to remove and/or separate species of that type. The first type of separation requires extremely high selectivity for a particular species. Often, a reasonably selective separation system can be found for separating such a species at higher concentrations, but difficulties are encountered when it is dilute relative to other matrix elements. The high selectivity that macrocycles show for cations offers promise for the separation of these dilute elements. The second type of separations problem requires a separation technique in which the affinity of the species of interest for the separating molecule exceeds that of any matrix components for that species. Again, the use of molecules capable of molecular recognition as carriers in membrane systems should have potential for solving such problems due to the extremely strong interactions of such molecules with particular species as exemplified by high log $\underline{K}$ values for macrocycle-cation interactions. The combination of strong interaction and high host-guest selectivity characteristics makes such systems ideal candidates for separations systems. The third type of separations problem arises because standard separations methods are based on differences in physical and/or chemical properties, such as boiling point, species charge, hydrophobicity, etc. When the separation of only the species with the weakest of such properties in a particular matrix is desired, the other matrix elements must first be separated out. This is often not economical, and/or the economics for such a separation would be enhanced by the ability to simply remove the desired species in a single and highly selective separation step. An example of such a problem is when ion exchange resins are used to separate a monovalent cation such as $Ag^+$ from a matrix containing multivalent cations. In this case, the use of a molecular recognition carrier molecule such as a triazolo macrocycle in a membrane with high and selective affinity for $Ag^+$ could provide an appropriate solution.

A final advantage of membrane systems which use molecular recognition-based carrier molecules to perform separations is that such systems contain several fundamental design parameters which may be adjusted in a synergistic manner to obtain extremely high selectivities and product concentration factors. In the case of cation separations using macrocycles as carriers in membrane systems, one can choose to vary the macrocycle cavity size; choose between a neutral and proton-ionizable macrocycle; choose particular macrocycle donor atom number, type, and constituent group; choose the number of proton-ionizable sites in a carrier molecule; vary the membrane system receiving phase composition with acids, complexing agents, or other reagents; and alter the phase composition. This large number of choices affords the opportunity to design a separation system to fill the specific needs of a particular process. In addition, these membrane systems offer promise for the design of energy-efficient and easily and continuously engineered separation processes.

Future applications of membrane technologies involving the design of molecular selectivity into the systems require the following capabilities. First, the reactions occurring must be understood at the molecular level. Second, equilibrium constant data must be available for all reactions. Third, the synthetic capability must be available to prepare carrier molecules as needed.

## B. POTENTIAL AREAS OF APPLICATION

The primary application areas where the separations capability of carrier-mediated liquid membrane systems show potential are in the removal of unwanted species from environmental samples, analytical separation and/or concentration procedures, and industrial separations. Industrially, highly selective separations processes which could use molecular recognition technology include gas separations; precious metal purification, recovery, concentration, and refining; chiral molecule separations in the pharmaceutical and other industries; and other synthetic organic molecule separations. There are many molecules which need to be selectively concentrated to improve analytical detection limits as well as separated from interferent matrices for subsequent analysis. Molecular recognition technology coupled with appropriate membrane systems hold promise in solving specific problems in these areas. There are also many instances where industrial waste separations as well as potable water and other systems with environmental separation needs present problems possibly suited to be solved using carrier molecules in liquid membrane systems. This may be particularly true when low levels of certain toxic species need to be removed from matrices containing many other more concentrated nontoxic elements. In these situations, high selectivity, strong interactions, and concentrative ability are essential components of any process devised to handle the situation.

In conclusion, both the macrocyclic carrier molecules and other system parameters in emulsion and hollow fiber liquid membrane systems have been designed to perform highly selective cation separations. An understanding of all the system parameters which affect membrane performance is necessary to design a stable membrane system as well as to perform the desired separation. The use of macrocycles in membrane systems to separate cations illustrates the possibilities inherent in the use of molecular recognition technology in future separation systems.

## ACKNOWLEDGMENTS

Appreciation for financial support of the research upon which this article is based is expressed to the U.S. Department of Energy, Office of Basic Energy Sciences through Grant No. DE-FG02-86ER13463; Serpentix Conveyor Corporation, Westminster, Colorado; and the State of Utah Centers of Excellence Program. We also appreciate the research efforts of our many student and post-doctoral colleagues over the past several years.

## REFERENCES

1. **Danesi, P. R., Reichley-Yinger, L., and Tickert, P. G.,** Lifetime of supported liquid membranes: the influence of interfacial properties, chemical composition and water transport on the long term stability of the membranes, *J. Membr. Sci.,* 31, 117, 1987.
2. **Lonsdale, H. K.,** The growth of membrane technology, *J. Membr. Sci.,* 10, 81, 1982.
3. **Way, J. D., Noble, R. D., Flynn, T. M. and Sloan, E. D.,** Liquid membrane transport: a survey, *J. Membr. Sci.,* 12, 239, 1982.
4. **Ohki, A., Hinoshita, H., Takagi, M., and Ueno, K.,** Transport of iron and cobalt complex ions through a liquid membrane mediated by methyltrioctylammonium ion with the aid of redox reaction, *Sep. Sci. Technol.,* 18, 969, 1983.

5. **Izatt, R. M., Clark, G. A., Bradshaw, J. S., and Lamb, J. D.,** Macrocycle-facilitated transport of ions in liquid membrane systems, *Sep. Purif. Methods,* 15, 21, 1986.
6. **Bhave, R. R. and Sirkar, K. K.,** Gas permeation and separation by aqueous membranes immobilized across the whole thickness or in a thin section of hydrophobic microporous celgard films, *J. Membr. Sci.,* 27, 41, 1986.
7. **Kim, J. I. and Stroeve, P.,** Mass transfer in separation devices with reactive hollow fibers, *Chem. Eng. Sci.,* 43, 247, 1988.
8. **Noble, R. D.,** Two-dimensional permeate transport with facilitated transport membranes, *Sep. Sci. Technol.,* 19, 469, 1984.
9. **Kajiyama, T., Kikuchi, H., Terada, I., Katayose, M., Takahara, A., and Shinkai, S.,** Permeation properties of polymer (liquid crystal) composite membrane, in *Curr. Top. Polym. Sci.,* Vol. II, Ottenbrite, R. M., Utracki, L. A., and Inoue, S., Eds., Hansen Publishers, Munich, 1987, chap. 6.6.
10. **Meares, P.,** Synthetic membranes and their applications: a survey of activity in the UK, April 1985, a report prepared for the Science and Engineering Research Council.
11. **Goddard, J. D.,** A fundamental model for carrier mediated energy transduction in membranes, *J. Phys. Chem.,* 89, 1825, 1985.
12. **Behr, J. P., Kirch, M., and Lehn, J. M.,** Carrier-mediated transport through bulk liquid membranes: dependence of transport rates and selectivity on carrier properties in diffusion-limited processes, *J. Am. Chem. Soc.,* 107, 241, 1985.
13. **Kimura, E., Dalimunte, C. A., Yamashita, A., and Machida, R.,** A proton-driven copper(II) ion pump with macrocyclic dioxotetraamine. A new type of carrier for solvent extraction of copper, *J. Chem. Soc., Chem. Commun.,* 1041, 1985.
14. **Bartsch, R. A., Kang, S. I., and Charewicz, W. A.,** Separation of metals by liquid surfactant membranes containing crown ether carboxylic acids, *J. Membr. Sci.,* 17, 97, 1984.
15. **Dulyea, L. M., Fyles, T. M., and Whitfield, D.,** Membrane-transport systems. 5. Coupled counter transport of alkaline-earth cations and protons, *Can. J. Chem.,* 62, 498, 1984.
16. **Izatt, R. M., LindH, G. C., Bruening, R. L., Huszthy, P., McDaniel, C. W., Bradshaw, J. S., and Christensen, J. J.,** Separation of silver from other metal cations using pyridone and triazole macrocycles in liquid membrane systems, *Anal. Chem.,* 60, 1694, 1988.
17. **Lamb, J. D., Bruening, R. L., Izatt, R. M., Hirashima, Y., Tse, P. K., and Christensen, J. J.,** Characterization of a supported liquid membrane for macrocycle-mediated selective cation transport, *J. Membr. Sci.,* 37, 13, 1988.
18. **Izatt, R. M., Roper, D. K., Bruening, R. L., and Lamb, J. D.,** Macrocycle-mediated cation transport using hollow fiber supported liquid membranes, *J. Membr. Sci.,* 45, 73, 1989.
19. **Izatt, R. M., Lamb, J. D., and Bruening, R. L.,** Comparison of bulk, emulsion, thin sheet supported, and hollow fiber supported liquid membranes in macrocycle-mediated cation separations, *Sep. Sci. Technol.,* 23, 1645, 1988.
20. **Izatt, R. M., Bradshaw, J. S., Nielsen, S. A., Lamb, J. D., Christensen, J. J., and Sen, D.,** Thermodynamic and kinetic data for cation-macrocycle interaction, *Chem. Rev.,* 85, 271, 1985.
21. **Kiggen, W. and Vögtle, F.,** More than two fold bridged phanes, in *Synthesis of Macrocycles: The Design of Selective Complexing Agents, Progress in Macrocyclic Chemistry,* Vol. 3, Izatt, R. M. and Christensen, J. J., Eds., John Wiley and Sons, New York, 1987, chap. 6.
22. **Weber, E.,** Crystalline uncharged-molecule inclusion compounds of unintended and designed macrocyclic hosts, in *Synthesis of Macrocycles: The Design of Selective Complexing Agents, Progress in Macrocyclic Chemistry,* Vol. 3, Izatt, R. M. and Christensen, J. J., Eds., John Wiley and Sons, New York, 1987, chap. 7.
23. **Vögtle, F., Sieger, H., and Müller, W. M.,** Complexation of uncharged molecules and crown-type host molecules, in *Host Guest Complex Chemistry. Macrocycles, Synthesis, Structures, Applications,* Vögtle, F. and Weber, E., Eds., Springer-Verlag, Berlin, 1985, 319.
24. **Hiraoka, M.,** Characteristics of crown compounds, in *Studies in Organic Chemistry. 12. Crown Compounds: Their Characteristics and Applications,* Hiraoka, M., Ed., Elsevier, Amsterdam, 1982, chap. 3.
25. **Lamb, J. D., Bruening, R. L., Linsley, D., Jones, C., and Izatt, R. M.,** Macrocycle mediated cation transport using two module hollow fiber supported liquid membranes, *Sep. Sci. Technol.,* in press.
26. **Izatt, R. M., Bruening, R. L., Bruening, M. L., LindH, G. C., and Christensen, J. J.,** Modeling diffusion-limited, macrocycle-mediated cation transport in supported liquid membranes, *Anal. Chem.,* 61, 1140, 1989.
27. **Izatt, R. M., Bruening, R. L., Bradshaw, J. S., Lamb, J. D., and Christensen, J. J.,** A quantitative description of macrocycle-mediated cation separation in liquid membranes and on silica gel as a function of system parameters, *Pure Appl. Chem.,* 60, 453, 1988.
28. **Izatt, R. M., Bruening, R. L., Bruening, M. L., and Lamb, J. D.,** Quantitating the effect of solvent type on neutral macrocycle-mediated cation transport in liquid membranes, *Isr. J. Chem.,* in press.
29. **Weast, R. C., Ed.,** *Handbook of Chemistry and Physics,* 67th ed., CRC Press, Boca Raton, FL, 1987.

30. **Stephen, H. and Stephen, T., Eds.,** *Solubilities of Inorganic and Organic Compounds,* Vol. 1, *Binary Systems,* Part 1, MacMillian, New York, 1963, 369, 377, 381, 476.
31. **Izatt, R. M., Bruening, R. L., Clark, G. A., Lamb, J. D., and Christensen, J. J.,** Effect of macrocycle type on $Pb^{2+}$ transport through an emulsion liquid membrane, *Sep. Sci. Technol.,* 22, 661, 1987.
32. **Lamb, J. D., King, J. E., Christensen, J. J., and Izatt, R. M.,** Determination of macrocyclic compounds in solution by thermometric titration against metal cations, *Anal. Chem.,* 53, 2127, 1981.
33. **Izatt, R. M., LindH, G. C., Bruening, R. L., Bradshaw, J. S., Lamb, J. D., and Christensen, J. J.,** Design of cation selectivity into liquid membrane systems using macrocyclic carriers, *Pure Appl. Chem.,* 58, 1453, 1986.
34. **Smith, R. M. and Martell, A. E.,** *Critical Stability Constants, Inorganic Complexes (Vol. 4), Amino Acids (Vol. 1),* Plenum Press, New York.
35. **Lamb, J. D., Christensen, J. J., Izatt, S. R., Bedke, K., Astin, M. S., and Izatt, R. M.,** Effects of salt concentration and anion on the rate of carrier-facilitated transport of metal cations through bulk liquid membranes containing crown ethers, *J. Am. Chem. Soc.,* 102, 3399, 1980.
36. **Izatt, R. M., Bruening, R. L., and Christensen, J. J.,** Use of co-anion type and concentration in macrocycle-facilitated metal cation separations with emulsion liquid membranes, in *Liquid Membranes: Theory and Applications,* Noble, R. D. and Way, J. D., Noble, R. D. V., Eds., American Chemical Society Press, Washington, D.C., 1987, 98.

Chapter 7.2

# EXCITABLE LIQUID MEMBRANES

## Kenichi Yoshikawa

## TABLE OF CONTENTS

# I. INTRODUCTION

The excitable or oscillatory phenomenon is one of the central subjects in biological science. Various biological systems exhibit rhythmic phenomena, e.g., circadian rhythms, beating heart, breathing, developing embryo, oscillation of nervous cells, etc. Life is connected with rhythms in general. In relation to the rhythmic phenomena in biological systems, there has been increasing interest in oscillatory chemical reactions in homogeneous media, espeically the Belousov-Zhabotinskii reaction.[1] The Belousov-Zhabotinskii reaction shows various interesting phenomena, e.g., limit-cycle oscillations, multi-periodicity, hysteresis, entrainment, chaos, and formation of spatial structures such as propagating target and/or spiral patterns. The driving force to produce these dissipative spatio-temporal structures is the oxidative chemical reaction of organic substrates such as malonic acid or citric acid. From the phenomenological point of view, the mechanism for this reaction is similar to that exhibited by living organisms, i.e., life on earth exists with the expense of chemical energy. Thus, the Belousov-Zhabotinskii reaction serves as a good model for understanding the mechanism of self-organization in living organisms.[2] The Belousov-Zhabotinskii reaction creates the spatio-temporal structures in a homogeneous solution. Conversely, living cells maintain their lives by use of cell-membranes, i.e., the interface or membrane plays an essential role in inducing rhythms or oscillations in biological systems. For example, electrical excitation in nervous cells is caused by the gating flow of sodium and potassium ions across the membrane.[3] The difference in the chemical potential of these ions is the driving force of the excitable phenomenon. Studies of artificial membranes which possess excitability are, therefore, important for understanding the mechanism of excitation and/or oscillation in biological systems.

Table 1 shows the history of the studies on oscillatory phenomena in artificial membranes.[4] Most of these studies have been carried out with the aim of increasing our understanding of the mechanism of oscillations in biomembranes. Our research group has also re-examined almost all of the experimental systems described in Table 1. We have noticed that the main problem in these studies is poor reproducibility, i.e., the occurrence of the oscillations was recorded, but details of the frequency, amplitude, and the pattern of the oscillations were rarely reported in the quantitative manner.

Recently, we found that sustained rhythmic oscillations are generated in a liquid membrane consisting of water/oil phases, where one of the aqueous phases contains a cationic[5-8] or anionic[9-11] surfactant. As the reproducibility of the oscillations in this liquid membrane is excellent, one can quantitatively study the effect of various chemical species on them. In the present article, we would like to describe how the frequency, amplitude, and shape of the electrical oscillations change markedly with the addition of various chemical species to the aqueous phase in the liquid membrane. Explanation of the mechanism of oscillation is given in connection with the nonlinear characteristics of the surfactant molecules at the interface. In the last section, a new sensing system with an external oscillatory circuit is proposed for detecting various chemical species. The basis for this system lies in one's ability to extract quantitatively the nonlinear characteristics of electrochemical systems.

# II. CHEMICAL SENSING WITH AN EXCITABLE LIQUID MEMBRANE

## A. OSCILLATORY PHENOMENA AT AN OIL/WATER INTERFACE

Under certain nonequilibrium conditions, self-movement of the interface is known to be generated in an oil-water system and to be accompanied by mass-transfer across the interface. Such a phenomenon is called the Marangoni effect[12] and is well-known among chemical engineers. In general, the movement of the interface arising from this Marangoni

## TABLE 1
## Oscillations in Artificial Membranes

| Membrane | Oscillation | Driving force | Researcher (year) |
|---|---|---|---|
| Glass filter | 1 V, 30min | Hydrostatic pressure (25 mA) | Teorell (1958) |
| Ion-exchange membrane | 0.1 V, 0.1sec; 0.1 V, 20msec | Direct current | Yamamoto (1961), Shashoua (1967) |
| Membrane from dry-oil | 0.1 V, 0.2sec | Direct current | Monnier (1964) |
| Bilayer lipid membrane + Proteneous compound | +20, 0, -20 mV, 1 sec | Direct current | Mueller and Rudin (1967) |
| Porous filter with a lipid | 20mV, 1 min | Direct current | Kobatake (1970) |
| Bilayer lipid membrane (BLM) | 10mV, 1 min | Chemical reaction Kl, pH 10 and $Fe(CN)_6^{3+}$, pH 5 | Pant and Rosenberg (1971) |
| Membrane with papain | $\Delta$ pH 0.2, 1 sec | Enzymic chemical reaction | Naparstek and Caplan (1973) |
| Oil-water interface | 400 300 200 100 mV, 5 sec | Concentration difference | Dupeyrat (1978), Yoshikawa (1983) |
| Porous filter with lipids | 25 20 mV, 5 sec | Concentration gradient | Yoshikawa and Ishii (1984) |

instability is depressed in the presence of surfactant molecules. In contrast to this effect, Dupeyrat and Nakache found quasi-periodic variations of a relaxation type in the interfacial tension and the electrical potential in an oil/water system in the presence of a cationic surfactant.[13,14] Recently, we have found that sustained rhythmic oscillations are generated in a liquid membrane consisting of water/oil/water phases, where one of the aqueous phases contains a cationic surfactant[5-8] or an anionic surfactant[9-11] or a phospholipid.[14] As the reproducibility of the oscillations in this liquid membrane is excellent, we could quantitatively

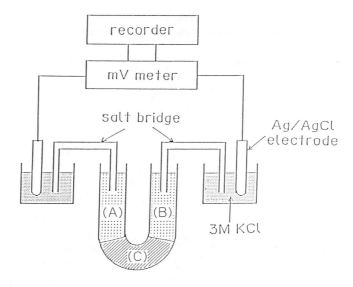

FIGURE 1.   Experimental apparatus for measurement of the oscillation of the electrical potential in a liquid membrane, water-oil-water system. (A) aqueous solution of surfactant containing alcohol, (B) 0.5 $M$ NaCl aqueous solution, and (C) nitrobenzene solution of 5 m$M$ 2,2'-bipyridine.

study the effect of various chemical species on the oscillations. In the present section, we describe the characteristic response of the oscillatory pattern in the liquid membrane to various chemical compounds.

## B.   EFFECT OF ALCOHOLS ON THE OSCILLATIONS[4,11,15]

Experiments were performed in an apparatus in a U-shaped glass tube (12 mm inner diameter). The apparatus is shown schematically in Figure 1. A nitrobenzene solution (4 ml) of 5 m$M$ 2,2'-bipyridine was placed in the base of the U cell. 2,2'-bipyridine was used to reduce the electrical resistance of the bulk organic phase and could be replaced by other hydrophobic ions, such as the tetrabutylammonium ion. Aqueous solutions (10 ml in each arm), one of which contained the surfactant, were introduced simultaneously into the arms of the U cell above the organic phase, without stirring. All measurements were carried out at 25°C. The voltage across the liquid membrane was measured with a high-impedance voltmeter connected by two salt-bridges to two Ag/AgCl electrodes. Figure 2 shows the tracing of the voltage oscillations across the organic phase of nitrobenzene in the presence or absence of soybean lecithin and/or butanol. Only small fluctuations of the electrical potential were observed with soybean lecithin (Figure 2A), and somewhat irregular oscillations were generated with butanol (Figure 2B). Rhythmic, large pulses were observed in the presence of both soybean lecithin and butanol (Figure 2C). These results apparently suggest that the transfer of the surfactant and alcohol through the oil/water interface is essentially important for inducing the rhythmic oscillations. It should be noted that rhythmic and sustained oscillations are generated spontaneously in the absence of any external stimuli, such as hydrostatic pressure or electrical current. In this liquid membrane, the concentration of the surfactant, together with the alcohol, is in "far-from-equilibrium" conditions. Thus, the dissipation of the free energy due to the solute transfer induces the oscillations.

We shall now demonstrate the characteristic effect of alcohol on the oscillatory pattern. The changes in patterns of oscillations of the liquid membrane with CTAB (cetyltrimethyl ammonium bromide, a cationic surfactant), SDS (sodium dodecyl sulfate, an anionic surfactant), and sodium oleate (a soap), in response to added alcohols, are shown schematically

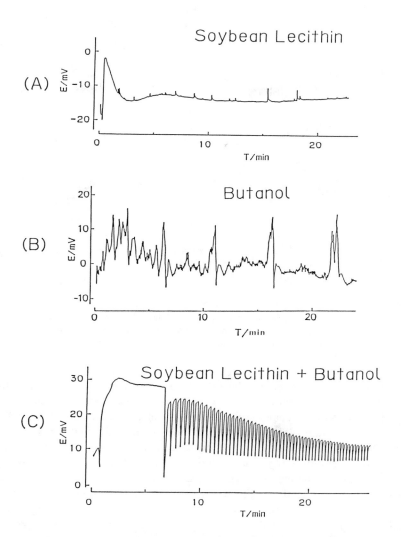

FIGURE 2. Oscillation of the electrical potential with (A) 0.0035 w/v % soybean lecithin, (B) 6 v/v % 1-butanol, and (C) 6 v/v % 1-butanol and 0.0035 w/v % soybean lecithin.

in Figure 3.[4] In the liquid membrane with CTAB, the frequency increased with an increase in the hydrophobicity of the alcohol, whereas the amplitude remained essentially constant. For the liquid membrane with SDS, the shape of the oscillations changed remarkably. In the liquid membrane with sodium oleate, the amplitude increased with the increase of the size of the alkyl chain of the alcohol.

When the concentration of the alcohol was increased, the frequency increased in the liquid membrane with CTAB. A similar change of the oscillations was encountered for the liquid membrane with SDS. However, no marked change of oscillation was observed for the liquid membrane with sodium oleate. These results suggest that, by studying the excitable phenomena in various liquid membranes, it may be possible to develop a new type of chemical sensor capable of distinguishing various chemical substances quantitatively, using information based on the amplitude, frequency, and shape of electrical pulses.[6]

## C. TASTE-PROFILE OF THE LIQUID MEMBRANE[11,15]

We have examined the effect of chemical substances belonging to the four basic taste categories on the manner of the electrical oscillation. We used the liquid membrane of

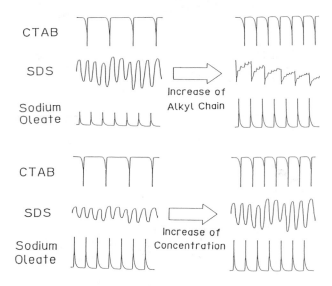

FIGURE 3.   Schematic representation of the effect on the manner of oscillation with (top) the increase of alkyl chain length and (bottom) the increase in concentration of alcohol in the liquid membrane with CTAB, SDS, and sodium oleate.

nitrobenzene between two aqueous phases, one of which contained 0.1 m$M$ sodium oleate and 10 v/v% 1-propanol. Approximately 1 h after the construction of the liquid membrane, when the oscillations had become stable, various chemical stimuli were added to the aqueous phase containing sodium oleate. Sodium chloride, sucrose, quinine chloride, and hydrochloric acid were examined as the typical substances of taste: salty, sweet, bitter, and sour, respectively. In the presence of sodium chloride, small rhythmic oscillations were generated together with giant oscillations of longer period, see Figure 4(A). When sucrose was added to the aqueous phase, chaotic oscillations were observed, as shown in Figure 4(D). It is quite interesting to note that the oscillatory characteristics, such as the amplitude, the frequency of modulation, and the shape of the pulses, change remarkably, depending on the nature of the chemical stimulant. It has been confirmed that other chemical stimulants for taste sensation also induce similar characteristic effects on the behavior of the oscillation, their nature being dependent upon the individual taste-categories, as shown in Figure 4.

# III. ON THE MECHANISM OF OSCILLATION

## A. SIMULTANEOUS MEASUREMENT OF ELECTRICAL POTENTIAL AND INTERFACIAL TENSION

Figure 5 shows the experimental apparatus for the simultaneous measurement of the electrical potential and the interfacial tension at the oil-water interface, using the Wilhelmy method. Figure 6 demonstrates that the periodic change of electrical potential synchronizes with the change of the interfacial tension for the liquid membrane in the presence of sodium oleate. As the interfacial tension is directly related to the concentration of surfactant molecules at the interface, the rhythmic change of the interfacial tension clearly indicates that the concentration of oleate at the oil/water interface changes repeatedly between high and low values. An increase in the concentration of oleate at the interface corresponds to the growth of the monolayer, which, in turn, increases the electrical potential of the electronic bilayer. The synchronized change of the electrical potential and the interfacial tension is thus explained.

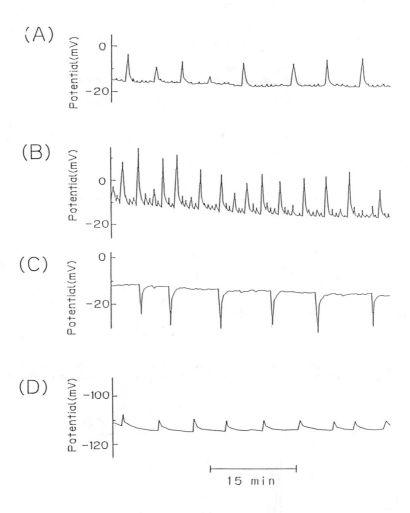

FIGURE 4. Oscillations of electrical potential, after addition of 0.5 ml of (A) 0.1 $M$ sodium chloride, (B) 1 $M$ sucrose, (C) 1 m$M$ quinine chloride, and (D) 30 m$M$ hydrochloride acid to the 10 ml aqueous solution of 10 v/v % 1-propanol and 0.1 m$M$ sodium oleate in the left hand cell.

## B. THEORETICAL INTERPRETATION OF THE OSCILLATIONS[11,15]

Based on the experimental results reported above, the mechanism for the rhythmic oscillations in the liquid membrane with sodium oleate and alcohol may be explained in the following way. The discussion is also valid with the oscillatory phenomena observed in the liquid membranes with the other surfactants. At a first step, oleate anions, which are mainly present as micelles in the aqueous phase, move towards the interface and become situated there on. The rates of migration of oleate and alcohol affect each other in a synergetic manner. Thus, the concentrations of oleate anion and alcohol at the interface increase gradually, and the surfactant, oleate, tends to form a monolayer structure at the interface. When the concentration of the anionic surfactant at the interface reaches an upper critical value, oleate anions are abruptly transferred to the organic phase with the formation of inverted micelles or inverted microemulsions. This step should be associated with the transfer of alcohol from the interface to the organic phase. The formation of inverted oleate-micelles may be assisted by the formation of acid-base pairs with an amine present in the organic phase. Thus, when the concentration of oleate at the interface decreases below the lower

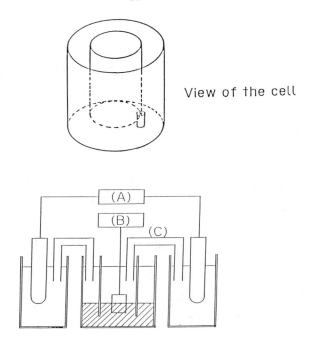

View of the cell

FIGURE 5.   Experimental apparatus for the simultaneous measurement of the electrical potential and interfacial tension at the oil-water interface: (A) m$M$ meter, (B) electronic balance, and (C) KCl salt bridge.

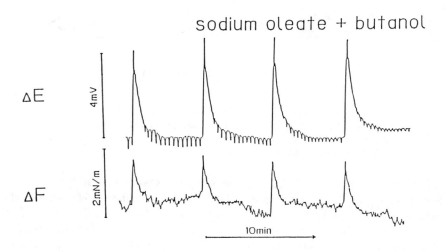

FIGURE 6.   Result of the simultaneous measurement of the electrical potential and interfacial tension at the interface between the aqueous solution of 0.01 m$M$ sodium oleate and 5 v/v % butanol and the nitrobenzene solution of 0.5 m$M$ tetraethylammonium bromide, in the apparatus shown in Figure 5.

critical value, accumulation of the surfactant begins again and the cycle is repeated. Changes in the concentration of oleate at the interface will be associated with the oscillation of the electrical potential, as has been demonstrated from the simultaneous measurement of the electrical potential and the interfacial tension. In other words, when an electrical double layer is formed around the interface with an increase in the concentration of oleate, it causes an interfacial negative electrical potential on the organic phase with respect to the aqueous phase. With the destruction of the monolayer, the interfacial potential is diminished.

The foregoing interpretation of the mechanism, although necessarily somewhat specu-
lative, provides a useful kinetic model. Let X, Y, and Z be the concentrations of the key
chemicals; $X_i$, the concentration of oleate at and/or near the interface; $Y_i$, the concentration
of alcohol near the interface; $Z_i$ the concentration of the aggregate or complex of oleate and
alcohol at and/or near the interface. Scheme 1 can be considered as a possible explanation
for the mechanism of oscillation in a liquid membrane. $X_b$ and $Y_b$ are the concentrations of
oleate and alcohol, respectively, in the bulk aqueous phase.

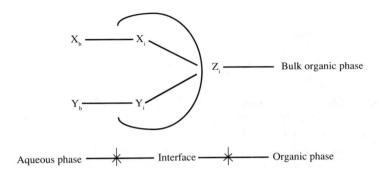

Scheme 1.    The process of diffusion of oleate and alcohol from the aqueous phase to the organic phase through the
interface.

Scheme 1 is composed of the following steps (i)-(iv), whose quantitative relationship
is not given for the sake of simplicity.

$$X_b \xrightarrow{D_X} X_i \qquad \text{(i)}$$

$$Y_b \xrightarrow{D_Y} Y_i \qquad \text{(ii)}$$

$$X_i + Y_i \xrightarrow{k_3} Z_i \qquad \text{(iii)}$$

$$Z_i \xrightarrow{k_4} \text{bulk organic phase} \quad \text{(iv)}$$

Steps (i) and (ii) correspond to the diffusion of oleate and alcohol from the bulk aqueous
phase to the interface. Step (iii) is the formation of aggregates of oleate and alcohol at the
interface, and is related to the construction of the monolayer associated with oleate and
alcohol. Step (iv) indicates the migration of the aggregates of oleate and alcohol from the
formation of interface to the bulk organic phase, a process which leads to the formation of
inverted micelles or microemulsions in the organic phase.

It may be expected that the rates of transfer of oleate and alcohol, $D_X$ and $D_Y$, are
affected by the negative feedback of $Z_i$. In other words, the diffusion rate, $D_X$, of oleate
from the aqueous phase to the interface may decrease with the net increase in the concentration
of oleate at the interface, $X_i$ plus $Z_i$. A similar situation may hold for the diffusion rate,
$D_Y$, of alcohol from the aqueous phase to the interface. Hence, the system kinetics may be
considered under the following assumptions: (a) the concentrations of oleate and alcohol in
the bulk aqueous phase, $X_b$ and $Y_b$, remain constant; (b) the rates of diffusion of oleate and
alcohol from the bulk aqueous phase to the interface are expressed as $D_X(X_b - X_i)$ and
$D_Y(Y_b - Y_i)$, respectively; (c) the negative feedbacks of $Z_i$ on the diffusion of X and Y are
given by $-k_1 Z_i$ and $-k_2 Z_i$, respectively; (d) the rate of step (iv) is expressed as a function,

$F(X_i,Y_i)$, of the rate constant $k_3$; and (e) the rate of step (iv) is expressed as a function, $G(Z_i)$, of the rate constant $k_4$. Under these assumptions, the kinetics of the migration of oleate and alcohol are written by the differential equations:

$$\frac{dX_i}{dt} = D_X(X_b - X_i) - k_1Z_i \tag{1-a}$$

$$\frac{dY_i}{dt} = D_Y(Y_b - Y_i) - k_2Z_i \tag{1-b}$$

$$\frac{dZ_i}{dt} = k_3F(X_i,Y_i) - k_4G(Z_i) \tag{1-c}$$

$F(X_i,Y_i)$ is given by $c(X_i + Y_i)$, a simple form for the synergetic effect of $X_i$ and $Y_i$, which has the physical meaning that the monolayer is formed together with oleate and alcohol. Self-oscillatory states can be obtained if $G(Z_i)$ has "N"-shape nonlinearity.

According to the two-dimensional van der Waals' equation for the surfactant molecules at the interface, an "N"-shape relationship is expected for a plot of the surfactant pressure, $\Pi$, against concentration of surfactant molecules, $\Gamma$.[15] We estimate that the rate, $k_4G(Z_i)$, is proportional to the surface pressure, $\Pi$. Further discussion on this "N"-shape nonlinearity will be given in the following section.

For the convenience of numerical analysis it is convenient to redefine the variables thus:

$$x = X_i, \quad y = Y_i, \quad z = Z_i \tag{2}$$

Then Equations (1-a to 1-c) become:

$$dx/dt = k_1(a_1 - b_1x - z) \tag{3-a}$$

$$dy/dt = k_2(a_2 - b_2y - z) \tag{3-b}$$

$$dz/dt = k_3[c(x + y) + G(z)] \tag{3-c}$$

In order to simplify the analysis it is also assumed that $k_3$ equals $k_4$.

Figure 7 exemplifies the numerical results of the equations. Only the parameter $a_1$ and the "N"-shape function, $G(z)$, are changed, and the other parameters, $k_1$, $k_2$, $k_3$, $b_1$, $a_2$, $b_2$ and $c$, are constant. On the right hand side of Figure 7, the function $G(z)$ is shown together with the straight line of the first-order rate equation, (1-a), which has been calculated using linearized stability analysis. It is quite interesting to note that the numerical results in Figure 7(A), (B), (C), and (D) correspond well to those of the experimental trends shown in Figure 4, suggesting that the characteristic change of the manner of oscillation is attributable, for the most part, to the change in the rate of diffusion of oleate and/or alcohol and also to the change in the state of the monolayer. The change of the diffusion rate, corresponding to the change in the parameters $k_1$ and $k_2$, may be induced by the chemical substances added to the aqueous phase. The stability of the monolayer, corresponding to the change of function $G(z)$, may also be affected, in the different manners, by the chemical stimuli. Though the experimental foundation of the parameters has not been satisfied to date, it has been shown that the characteristic manner of the oscilliations in the liquid membrane can be interpreted using the above kinetic equations.

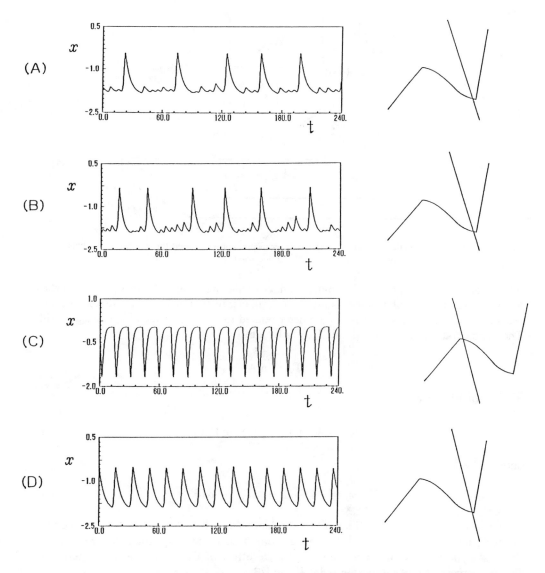

FIGURE 7. Computer simulation of the electrical oscillation. To the right of the figure, the function G(z) (Equation (1-c)) is drawn together with the straight line of the first-order rate equation (Equation (1-a)) which has been calculated using linearized stability analysis.[2]

## IV. NONLINEAR CHARACTERISTICS OF A SURFACTANT MONOLAYER

### A. ON THE "N"-SHAPE NONLINEARITY IN THE RATE EQUATION

In Equation 3, the nonlinearity of the "N"-shape function, $G(z_i)$, plays an essential role in obtaining various oscillatory regimes by theoretical analysis. In general, such an "N"-shape relationship is correlated with various oscillatory phenomena in far-from-equilibrium conditions. Therefore, we now discuss the physico-chemical basis of the nonlinearity in our experimental system. Let us start from the two-dimensional van der Waals' equation,[16] which expresses the behavior of the surfactant molecules at the oil-water interface,

$$[\pi + \alpha(n/A)^2](A - n\beta) = nRT \tag{4}$$

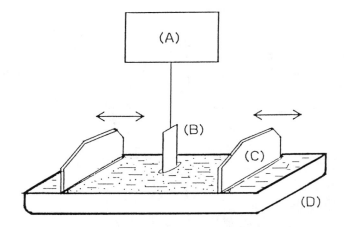

FIGURE 8.    Schematic representation of the experimental apparatus used for measuring the dynamic interfacial tension of a lipid monolayer. (A) Electronic balance, (B) Pt plate, (C) barrier, (D) trough.

where, $\Pi$: interfacial pressure; n: mole number of the surfactant; A: area of the interface; R: gas constant; T: absolute temperature; $\alpha$: a parameter related to the attractive force between the surfactant molecules; and $\beta$: a parameter related to the size of the surfactant molecule. The concentration of the surfactant, $\Gamma_s$, is given by

$$\Gamma_s = n/A \tag{5}$$

Thus, Equation 5 becomes

$$(\pi + \alpha\Gamma_s^2)(1 - \beta\Gamma_s) = \Gamma_s RT \tag{6}$$

and can be rewritten as

$$\pi = RT\Gamma_s/(1 - \beta\Gamma_s) - \alpha\Gamma_s^2 \tag{7}$$

Equation 7 indicates that the functional relationship, $\Pi$ vs. $\Gamma_s$, has an "N"-shape. That is, as $\Gamma_s$ increases from zero to the critical point $1/\beta$, $\Gamma$ increases gradually from zero by the first hyperbolic term, then decreases by the effect of the second quadratic term, and finally increases to infinity at the point $\Gamma_s = 1/\beta$.

With the increase of the interfacial pressure, $\Pi$, the rate of migration, $G(Z_i)$, of the aggregate of oleate and alcohol from the interface to the bulk organic phase, may increase. The rate, $G(Z_i)$, is thus expected to possess "N"-shape characteristics. Although this scheme may be too simple to interpret the full detail of the oscillatory behavior in the liquid membrane, the above discussion may shed new light on this interesting phenomenon. Further theoretical studies concerning the oscillatory mechanism are awaited, together with measurements of the absolute rates of the transfer of oleate and alcohol.

## B. DYNAMIC SURFACE TENSION AS AN INDEX OF NONLINEARITY

In order to ascertain the characteristic nonlinear behavior of surfactant molecules, we have measured the dynamic surface tension for the monolayer of oleic acid at an air/water interface. Figure 8 shows the experimental apparatus used for measuring dynamic surface tension. A monolayer of oleic acid was spread on the aqueous phase of water purified by Milli-Q filtration after double distillation. The moving teflon trough changed the surface area repeatedly with a sinusoidal function. The change of the surface pressure was contin-

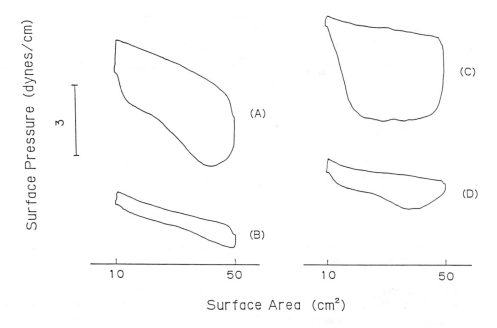

FIGURE 9. Dynamic surface tension for the monolayer of oleic acid in the presence of (A) 0.1 $M$ sodium chloride, (B) 1 $M$ sucrose, (C) 1 m$M$ quinine chloride, and (D) 30 m$M$ hydrochloric acid.

uously monitored with an electronic balance. Figure 9 shows the dynamic surface tension for the monolayer of oleic acid in the presence of (A) 0.1 $M$ sodium chloride, (B) 1 $M$ sucrose, (C) 1 m$M$ quinine chloride, and (D) 30 m$M$ hydrochloric acid.

Similar characteristic changes of the dynamic surface tension were observed, depending on the category of taste: salty, sweet, bitter, and sour. Inorganic salts, as the salty compounds, may affect the Debye length of the change effect of the surfactant head-group and may change the manner of aggregation of the surfactant molecules. Sweet compounds, such as sucrose, change the dynamic structure of water molecules, which, in turn, affects the manner of aggregation of the surfactants. Bitter compounds are generally rather hydrophobic and are expected to change the manner of packing of alkyl-chains in the aggregates of the surfactant molecules.[17] Sour compounds change the degree of dissociation of the head-group of the surfactant molecules. It therefore becomes apparent that each compound, belonging to a different taste category, affects the dynamic interfacial tension or the nonlinear characteristics of the $\Pi$, $\Gamma$ relationship. This phenomenon is expected to be related to the characteristic response of the liquid membrane to the various taste-stimuli, as shown in Figure 4, in spite of differences in the experimental systems, i.e., the dynamic interfacial tensions were obtained for the air/water interface whereas the oscillatory phenomena were observed for the oil/water interface.

# V. A NOVEL CHEMICAL SENSOR WITH AN EXTERNAL ELECTRICAL OSCILLATOR

## A. METHODOLOGY

We have shown that the excitable liquid membrane can discriminate various chemical substances. However, the liquid membrane reported in this study presents some problems in its practical application. Therefore, in this section we shall describe a new sensing system, for detecting and quantitating various chemical species, which is based on the information of electrochemical nonlinearity.

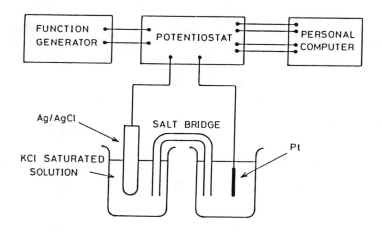

FIGURE 10.   Experimental apparatus used for measuring the nonlinear characteristics in an electrochemical system.

Various electrochemical sensors have been developed, such as a pH electrode and selective electrodes for inorganic-metal cations. In these sensors, chemical information, such as structure or activity, is converted into a DC voltage (or DC current) difference. As well as measuring the DC voltage or current, it may be useful to obtain chemical information from the capacitance or resistance of a test solution. In general, the capacitive component in an electrochemical system can be classified into two components: the dielectricity of the bulk of the solution and the capacitance on the electrode-surface originating from the electrical double layer of ion and electrosorption at the surface of the electrode.[18] Though there have been many physico-chemical studies on the capacitance in electrochemical systems, there have been few practical applications. Their absence may arise from the unsatisfactory reproducibility of the observed capacitance, caused by the hysteresis and variation of the surface area, including bubble formation on the electrode during its handling. It should be noted that the capacitance usually depends on the electrical potential applied to the electrode.[18]

Recently, we reported that, by using an external forced oscillatory circuit, "nonlinearity" of the electrochemical properties at the electrode-solution interface provides various useful information on the chemical substances present in solution.[19-21] However, in these studies the physico-chemical meaning of the detected "nonlinearity" remained unclear. In this discussion, we report a new chemical sensing system for detecting the "nonlinear" capacitive component. The procedure consists of monitoring the output-current under application of a sinusoidal voltage and then obtaining the relative intensities of its higher harmonics by Fourier-transformation, as shown in Scheme 2.

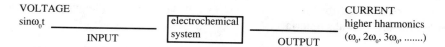

Scheme 2. Procedure for detecting nonlinear characteristics of an electrochemical system.

The experimental apparatus is shown schematically in Figure 10. The Pt-electrode (2 mm wire, 0.5 mm $\phi$) was immersed in the test solution (right hand cell in Figure 10) containing 10 m$M$ $MgCl_2$ and the surfactant (at various concentrations). $MgCl_2$ was added to the solution as a supporting electrolyte to reduce the impedance of the bulk aqueous solution. The test solution was connected by a 3 $M$ KCl agar bridge to a saturated KCl solution (left hand cell in Figure 10) in which was immersed an Ag/AgCl standard electrode.

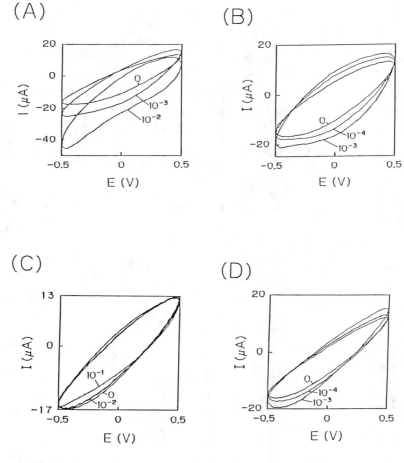

FIGURE 11. Current-voltage trace for the aqueous solutions, (A) hydrochloric acid, (B) ammonium chloride, (C) sucrose, and (D) nicotine. The numbers given in the figure are the concentrations of the chemical species.

Then, a purely sinusoidal voltage [amplitude: 1 volt (peak to peak); frequency: 1 Hz] generated by a function generator, was applied to the electrode through a potentiostat. The AC voltage was applied at a relatively low frequency (1 Hz), because the capacitive component due to the bulk dielectricity of the solution may be dominated at the higher frequencies. The output AC current was Fourier transformed into the frequency domain by use of a personal computer. All measurements were performed at $20 \pm 2°C$.

## B. DETECTION AND QUANTITATION OF CHEMICAL SPECIES

In the preceding section, we described how it is possible to distinguish chemical substances with the four basic taste categories from the differences in patterns of self-oscillation in a water-oil-water liquid membrane containing sodium oleate. It has also been shown that the dynamic $\Pi$, is a characteristic of an oleate-monolayer change, depending on the chemical species added to the aqueous layer. We have, therefore, studied the change of the electrochemical characteristics of a solution of sodium oleate after addition of various taste-compounds, belonging to the four basic taste-categories: sweet, sour, salty, and bitter, while making a special effort to detect quantitatively any nonlinearity in the system.

Figure 11 exemplifies the change of the I-V (I: current, V: sinusoidal voltage of 1 Hz) curve in a 1 m$M$ sodium oleate aqueous solution caused by the addition of the following

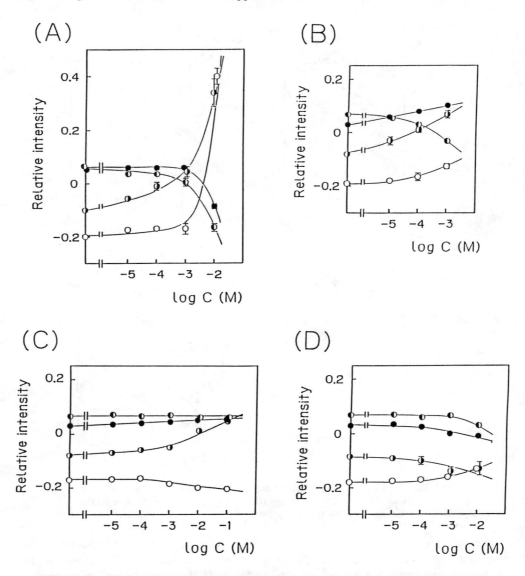

**FIGURE 12.**    Fourier-transformation of the output current with the application of a sinusoidal voltage. The experimental condition are the same as those in Figure 11.

taste compounds, (A) hydrochloric acid as a sour compound, (B) ammonium chloride as a salty compound, (C) sucrose as a sweet compound, and (D) nicotine as a bitter compound. It is interesting to note that the effects of the taste-compounds are apparently different from each other. In order to analyze the characteristics of the I-V curve quantitatively, the time-trace of the output current was Fourier-transformed to the frequency domain (Figure 12). $R_n$ and $I_n$ correspond to the intensities of the real part (conductance) and the imaginary part (capacitance) of the $n_{th}$ harmonics, respectively. It was found that the relative intensities characteristically change with the concentration of the taste compounds. In order to minimize

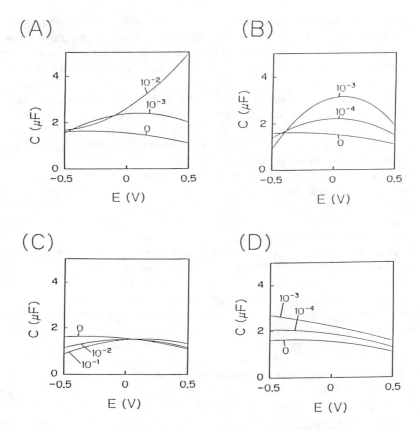

FIGURE 13. Simulated capacitance-voltage relationship based on the experimental results (Figure 12) of the Fourier-transformation of the output current. The experimental conditions are the same as those in Figure 11.

the experimental error, the intensities of the second and third harmonics are given as the relative value normalized to the imaginary component of the first harmonics, $I_1$, which correspond to the linearized capacitance with the neglect of nonlinearity. The experimental errors in the relative intensities were found to become less than 10% for all of the measurements in this study (Figure 12).

It has been found that the manner of the response clearly depends on the category in which the taste-compounds belong. We have examined at least three compounds for each category: for example, for sour compounds, acetic acid, tartaric acid, citric acid; for salty compounds, NaCl, LiCl, KCl; for bitter compounds, theobromine, quinine hydrochloride, caffeine. From the results of the electrochemical measurements of these compounds, it became apparent that the sensing method reported herein is quite useful for distinguishing and quantitating various chemical substances, if one uses the information provided by the higher harmonics components.

## C. NONLINEARITY OF CAPACITANCE IN RELATION TO THE HIGHER HARMONICS

Figure 13 shows the simulated capacitance-voltage curve which is derived from the experimental results of the Fourier-transformation of the output current with the application of sinusoidal voltage. In this computer simulation, the components of the second and third harmonics have been evaluated using Equation 3 and the components of the higher harmonics, such as fourth and fifth harmonics, have been ignored. This figure clearly demonstrates that the nonlinearity of the capacitance changes in a characteristic manner after the addition of taste-compounds. It is important to note that such nonlinear behavior has rarely been obtained in a reproducible manner in classical impedance measurements because the observed values of capacitance generally exhibit time-dependent change and significant hysteresis. Let us now discuss the characteristic change of the nonlinearity of capacitance. Sour components, such as hydrochloric acid or tartaric acid, lower the pH value and induce a smaller degree of dissociation of the carboxylic acid. As a result, the charge density on the electrode, due to the anionic oleate adsorbed on the electrode surface, may be decreased. The change of the electron charge density on the electrode surface may change the shape of the voltage-dependent capacitance, as shown in Figure 13(A).

Salty compounds, such as sodium chloride or ammonium chloride, affect the electronic diffuse double layer around the surface of electrode. According to the Gouy-Chapman theory, which depends upon the assumption of the Boltzmann-distribution and the Maxwell-Planck relationship, the capacitance should increase with an increase in the concentration of the electrolytes. Figure 13(B) corresponds to this expectation.

The effect of the sweet compounds on the nonlinear characteristic of the capacitance is not as great but, nevertheless, a change is apparently induced. It is generally believed that sweet compounds change the structure of water clusters. These changes in the clusters may affect the manner of adsorption and/or aggregation of oleate molecules.

Bitter compounds, such as nicotine, are assumed to be incorporated mainly into the hydrophobic region of the aggregated lipids or surfactant molecules. It is, therefore, expected that the bitter compounds have the effect of increasing the adsorption of oleate onto the surface of the electrode. This effect may increase the capacitance, as seen in Figure 13(D).

At the present time, this interpretation of the nonlinear characteristics, which is dependent on the categories of taste compounds, is primitive. However, at least it is clear that the taste compounds, belonging to the different taste categories, affect the nonlinear properties of the electrode in different ways. With adequate modification of the surface of the electrode, the selective response of the nonlinearity to various chemicals may be improved. Studies along this line are awaited.

## VI. RELATIONSHIP OF THE CHARACTERISTIC RESPONSE OF THE ARTIFICIAL SYSTEM TO THE MECHANISM OF CHEMICAL SENSING IN LIVING ORGANISMS

Owing to the development of biochemical studies, many receptor proteins have been identified. It is quite apparent that living organisms distinguish between specific compounds by using these receptor proteins, e.g., pheromones in insects, antigen-antibody interactions, etc. On the other hand, there are many thousands of chemical compounds in the surroundings of the living organisms. Is it possible for living organisms to distinguish thousands of chemicals using thousands of selective receptor proteins? At least it seems that the selectivity to chemical substance of a single sensing cell is not high, but rather shows a broad spectrum.[22] Let us think of the stage of birth of initial life on earth. If it were a necessary condition that the initial cells had selective receptor proteins, living organisms would never have appeared. It is, therefore, expected that the initial lives on the earth were equipped with a simple chemical-sensing device as well as many selective receptor-proteins. The results of this manuscript, together with the previous studies in our laboratory, indicate that the manner of aggregation of oleate molecules changes characteristically and depends on the nature of added taste-compounds. Recently, we have found that phospholipids possessing an unsaturated alkyl moiety, such as the oleyl chain, can constitute an ion-channel without the benefit of a "channel-protein".[23] This fact also suggests that unsaturated lipids may have been important for living organisms in the initial stage of their long evolutionary process. Because the idea suggested in this article is novel and the theory to support it is still in the preliminary stage, further studies are awaited for its justification.

# ACKNOWLEDGMENTS

The author thanks Prof. H. Kawakami and Mr. T. Ishii for helpful suggestions. He is grateful to Mrs. S. Nakata, M. Makino, and N. Oyama.

# REFERENCES

1. **Field, R. J. and Burger, M.,** Eds., *Oscillations and Traveling Waves in Chemical Systems,* John Wiley and Sons, New York, 1985.
2. **Nicolis, G. and Prigogine, I.,** *Self-Organization in Nonequilibrium Systems,* John Wiley and Sons, New York, 1977.
3. **Hodgkin, A. L. and Huxley, A. F.,** Currents carried by sodium and potassium ions through the membrane of the Grant Axon of Loligo, *J. Physiol.,* 116, 449, 1952.
4. **Yoshikawa, K.,** Oscillatory phenomenon in chemistry-development of temporal order in molecular assembly, in *Dynamical Systems and Applications,* Aoki, N., Ed., World Scientific, Singapore, 1987, 205.
5. **Yoshikawa, K. and Matsubara, Y.,** Spontaneous oscillation of electrical potential across organic liquid membrane, *J. Am. Chem. Soc.,* 105, 5767, 1983.
6. **Yoshikawa, K. and Matsubara, Y.,** Chemoreception by an excitable liquid membrane: characteristic effects of alcohols on the frequency of electrical oscillation, *J. Am. Chem. Soc.,* 106, 4423, 1984.
7. **Yoshikawa, K., Omochi, T., and Matsubara, Y.,** Chemoreception of sugars by an excitable liquid membrane, *Biophys. Chem.,* 23, 211, 1986.
8. **Yoshikawa, K., Omochi, T., Matsubara, Y., and Kawakami, H.,** A possibility to recognize chirality an excitable artificial liquid membrane, *Biophys. Chem.,* 24, 111, 1986.
9. **Yoshikawa, K., Nakata, S., Omochi, T., and Colaccico, G.,** Novel liquid-membrane oscillator with anionic surfactant, *Langmuir,* 2, 715, 1986.
10. **Nakata, S., Yoshikawa, K., and Ishii, T.,** Excitable liquid membranes with anionic surfactant, *J. Chem. Soc. Jpn., Chem. Ind. Chem.,* (3), 495, 1987.
11. **Yoshikawa, K., Shoji, M., Nakata, S., and Maeda, S.,** An excitable liquid membrane possibly mimicking the sensing mechanism of taste, *Langmuir,* 4, 759, 1988.
12. **Søensen, T. S.,** Ed., *Dynamics and Instability of Fluid Interfaces,* Springer-Verlag, Berlin, 1979.
13. **Dupeyrat, M. and Nakache, E.,** Direct conversion of chemical energy into mechanical energy at an oil water interface, *Bioelectrochem. Bioenerg.,* 5, 134, 1978.
14. **Nakache, E., Dupeyrat, M., and Vignes-Adler, M.,** The contribution of chemistry to new marangoni mass-transfer instabilities at the oil/water interface, *Faraday Discuss. Chem. Soc.,* 77, 189, 1984.
15. **Yoshikawa, K., Maeda, S., and Kawakami, S.,** Various oscillatory regimes and bifurcations in a dynamical chemical system at an interface, *Ferroelectrics,* 77, 281, 1988.
16. **Adamson, A. W.,** *Physical Chemistry of Surfaces,* Interscience, New York, 1960.
17. **Kumazawa, T., Nomura, T., and Kurihara, K.,** Liposomes as model for taste cells: receptor sites for bitter substances including $N - C = S$ substances and mechanism of membrane potential changes, *Biochemistry,* 27, 1239, 1988.
18. **Bard, A. J. and Faulkner, L. R.,** *Electrochemical Methods,* John Wiley and Sons, New York, 1980.
19. **Yoshikawa, K. and Omochi, T.,** Chemical sensing by a novel electrical oscillator: detection and quantitation of polysaccharides in concanavalin a solution, *Biochem. Biophys. Res. Commun.,* 137, 978, 1986.
20. **Yoshikawa, K., Omochi, T., Fujimoto, T., and Terada, H.,** A novel method for the detection and measurement of inorganic cations from the frequency of electrical oscillations, *J. Coll. Interface Sci.,* 113, 585, 1986.
21. **Yoshikawa, K. and Ishii, T.,** Excitable Artificial membrane mimicking the biomembranes of nerve, taste and olfaction, Physical Organic Chemistry 1986, Kobayashi, M., Ed., *Studies in Organic Chemistry,* Vol. 31, Elsevier, Amsterdam, 1987, 477.
22. **Finger, T. E. and Silver, W. L.,** Eds., *Neurobiology of Taste and Smell,* John Wiley and Sons, New York, 1987.
23. **Yoshikawa, K., Fujimoto, T., Shimooka, T., Terada, H., Kumazawa, N., and Ishii, T.,** Electrical oscillation and fluctuation in phospholipid membranes-phospholipids can form a channel without protein, *Biophys. Chem.,* 29, 293, 1988.

Chapter 7.3

# LIQUID MEMBRANE TRANSPORT CONTROLLED BY RESPONSIVE FUNCTIONS

**Seiji Shinkai and Tsutomu Matsuda**

# TABLE OF CONTENTS

# I. INTRODUCTION

Iron is transported across a biomembrane by ionophores called "siderophores", the binding sites of which consist of either hydroxamic acids or catechols.[1] The stability constants of the ionophores are so large ($K_s$ = ca. $10^{20}$ to $10^{50} M^{-1}$) that it is easy for these ionophores to catch the iron into the biomembrane, but decomplexation by which the iron is released to the receiving aqueous phase scarcely occurs. The skillful method is realized in nature in order to effect the facile iron release. Thus, the strategy employed by the biological membrane transport system is either to reduce $Fe^{3+}$ (large $K_s$ with siderophores) to $Fe^{2+}$ (small $K_s$ with siderophores) or to decompose the siderophores by enzymes. The $Na^+$ transport method employed by monensin (Structure 1) seems more suggestive: it efficiently carries $Na^+$ from the basic aqueous phase to the acidic aqueous phase across the membrane.[2] At the ion-complexation side a pH-induced dissociation of the terminal carboxylate group enforces the cyclic structure due to the hydrogen-bonding with another terminal OH group, while at the acidic ion-release side the cyclic structure is broken down by neutralization of the carboxylate group.[2] As a result, $Na^+$ is transported very efficiently across the membrane.

Then, what kind of ion transport is expected when artificial ionophores such as crown ethers and cryptands are used as ion carriers? It is established that the best carrier for ion transport across a liquid membrane is an ionophore that gives a moderately stable rather than a very stable complex: that is, a plot of $K_s$ vs. transport rate results in a maximum at around $K_s = 10^6 M^{-1}$.[3-5] This means that the ion transport with simple ion carriers cannot exceed the maximal ceiling velocity because the ion-uptake from the source (IN) aqueous phase is a rate-limiting step when the $K_s$ is small, whereas the ion-release to the receiving (OUT) aqueous phase is a rate-limiting step when the $K_s$ is too large. The dilemma would be broken through, as the nature suggests us, if one can add a proper "switch function" to ion carriers by which the $K_s$ is changed in response to the stimuli from the outside world. The possible signals to induce these stimuli would be pH, light, heat, redox potentials, etc. In this chapter, we will review how the rate of carrier-mediated ion transport across liquid membranes is accelerated or controlled by these switch functions.

The oldest example for the switch-functionalized transport systems would be a pH-dependent system in which pH-responsive crown ethers bearing dissociable groups are used as ion carriers. These may be direct mimics of natural polyether antibiotics such as monensin, nigericin, lasalocid, etc. which have both a hydroxyl and a carboxyl group at the two sides of the polyether chain in order to effect the cyclic-noncyclic interconversion through the formation and scission of the hydrogen-bonds. Since the ion-transport system with pH-responsive crown ethers is described in Chapters 3, 4, and 6, we here survey other stimuli which can affect the complexation properties of a crown ether family.

# II. REDOX-SWITCHES

As described in the introduction, certain siderophores mediate $Fe^{3+}$ transport through the redox reactions. The ingenious $Cu^{2+}$ transport system with which the redox reaction is skillfully combined was designed by Matsuo et al.[6,7] Bathocuproine (Structure 2) strongly complexes $Cu^+$ which adopts the tetrahedral coordination, but cannot form the stable complex with $Cu^{2+}$ which adopts the square-planar coordination. Thus, $Cu^{2+}$ is extracted into the membrane phase only when $Cu^{2+}$ is reduced to $Cu^+$ by hydroxylamine.[6] At the ion-release side of the membrane, $Cu^+$ is readily oxidized to $Cu^{2+}$ by air resulting in a facile decomposition of the $Cu^+$-Structure 2 complex. The similar redox-assisted copper transport was achieved with tetradentate thioether (Structure 3).[7] It is known that the complex of 18-crown-6 with $Eu^{3+}$ has the low thermodynamic stability in aqueous solution because of the large solvation energy. Brown et al.[8] found that the successful transport of europium by 18-crown-

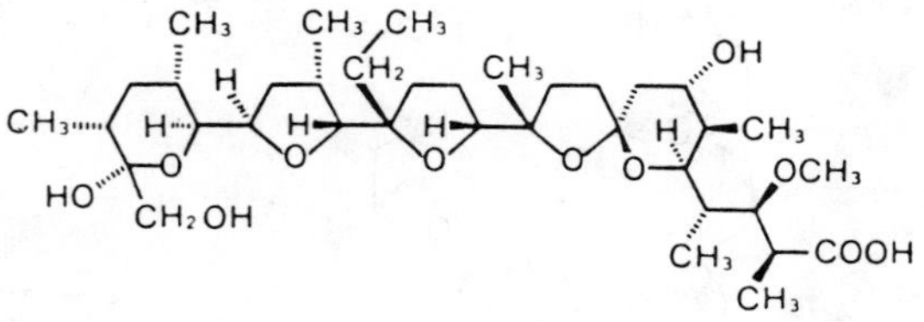

1

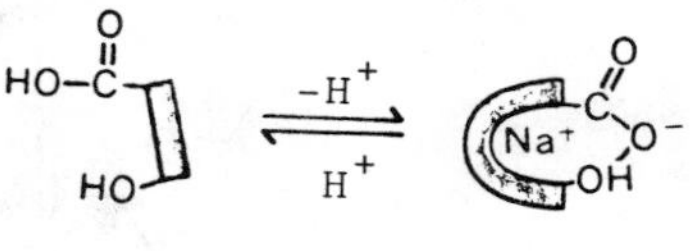

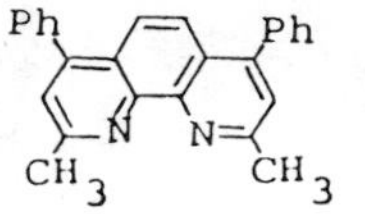

2

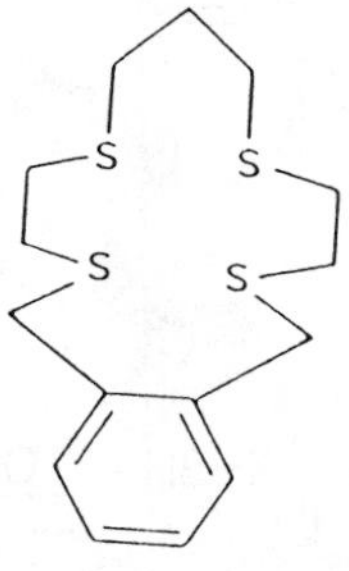

3

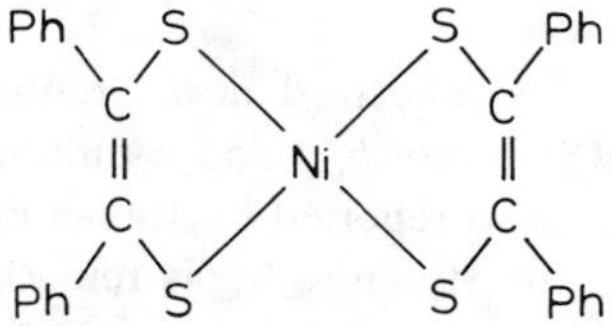

5

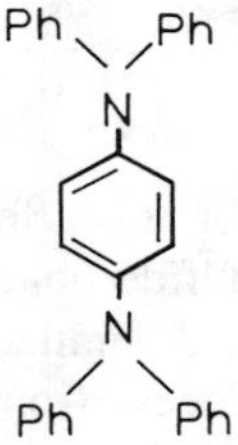

4

6 occurs when $Eu^{3+}$ is reduced to $Eu^{2+}$ which is subject to the relatively small solvation energy. Clearly, to change the $K_s$ by the redox of metal cations is a common idea in these reports.

The redox-control of the ion-transport rate is also achieved by the redox reaction of carriers. Two aqueous phases are bridged by a dichloroethane layer containing Structure 4: the IN aqueous phase contains $K_3Fe(CN)_6$ and the OUT aqueous phase contains ascorbic acid. It was found that picrate anion is transported from the IN to the OUT aqueous phase in an active transport manner.[9] At the IN interface Structure 4 is oxidized to Structure $4^+$ which serves as a carrier for picrate anion. At the OUT interface picrate anion is released to the OUT aqueous phase by virtue of a discharge of Structure $4^+$ to Structure 4 due to the reducing agent. Similarly, the redox reaction between Structure 5 and Structure $5^-$ is applicable to $K^+$ transport through the liquid membrane in the presence of dicyclohexyl-18-crown-6.[10]

Crown ethers which have a redox-active group within the same molecule are classified as redox-switched crown ethers. One can expect that (a) the ion-binding ability of the crown ether site can be controlled by the redox state of the prosthetic redox-active site, or on the contrary, (b) the redox potential of the redox-active site can be controlled by the metal binding to the prosthetic crown ether site. In case these two sites influence each other, one may consider that they ''conjugate'' each other. The most direct change induced by the redox reaction would be reversible bond formation and bond scission leading to the cyclic-noncyclic interconversion. The redox reaction of a thioldisulfide couple would be the most

$6_{ox}$                    $6_{red}$

$7_{red}$                    $7_{ox}$

suitable candidate for this. Shinkai et al.[11,12] synthesized new "redox-switched" crown ethers bearing a disulfide bond in the ring (Structure $6_{ox}$) and a dithiol group at its chain ends (Structure $6_{red}$). A similar approach was also reported by Raban et al.[13] It was found that the oxidation process from Structure $6_{red}$ to Structure $6_{ox}$ is remarkably subject to the metal template effect: the oxidation of Structure $6_{red}$ in the absence of the template metal gave the polymeric products, but the main products in the presence of $Cs^+$ were cyclic compounds. The stability constant for Structure $6_{ox}$ (320 $M^{-1}$ for $Cs^+$ in propylene carbonate) was smaller than that for monobenzo-21-crown-7 (2072 $M^{-1}$), but much greater than that for Structure $6_{red}$ (ca. 50 $M^{-1}$). In ion transport across a liquid membrane, Structure $6_{ox}$ carried $Cs^+$ 6.2 times faster than Structure $6_{red}$. It thus became possible to regulate the rate of $Cs^+$ transport by the redox reaction between Structure $6_{red}$ and Structure $6_{ox}$ in the membrane phase. In the liquid membrane transport system, Structure $6_{red}$ is oxidized to Structure $6_{ox}$ by the addition of $I_2$ leading to the rate acceleration, whereas Structure $6_{ox}$ is reduced to Structure $6_{red}$ by the addition of 1,4-butanedithiol leading to the rate suppression.

Another method to control the binding constants by a thiol redox-switch is an interconversion between monocrown and biscrown. This idea can be realized by using a thiol-containing crown ether Structure $7$[14]: that is Structure $7_{ox}$ is capable of forming an intramolecular 1:2 sandwich complex with a metal cation and therefore shows the higher selectivity toward large alkali metal cations. It was found that (i) the affinity of Structure $7_{ox}$ (X = none) for alkali metal cations is almost equal to that of monobenzo-15-crown-5, whereas Structure $7_{red}$ (X = none) has an affinity greater than Structure $7_{ox}$ (X = none) probably because of the electron-donating effect of the 4'-mercapto group toward the metal binding crown center and (ii) Structure $7_{ox}$ (X = $CH_2$) has an affinity for large alkali metal cations greater than Structure $7_{red}$ (X = $CH_2$) because of the cooperative action of the two crown rings to form 1:2 cation-crown sandwich-type complexes. The lack of the sandwich-type complexation in Structure $7_{ox}$ (X = none) was accounted for by the dihedral angle of the disulfide bond: i.e., (Z)-conformation of diphenyl disulfide is energetically unfavorable and the distance between the two crown rings is too short to sandwich a metal cation even though

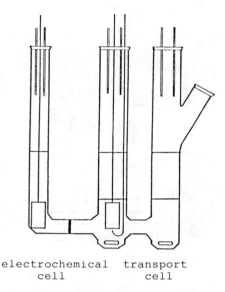

electrochemical    transport
        cell              cell

FIGURE 1.    Diagram of an electrochemically driven ion transport cell: (A) electrochemical cell, (B) liquid membrane cell.[17]

it adopts the (Z)-conformation. In contrast, Structure $7_{ox}$ (X = $CH_2$) has a $CH_2$ spacer suitable to the formation of the intramolecular sandwich complex. It was shown that in $K^+$ transport though the liquid membrane, the rate is efficiently accelerated when Structure $7_{red}$ (X = $CH_2$) is oxidized to Structure $7_{ox}$ (X = $CH_2$) by iodine added to the liquid membrane phase.

## III. ELECTROCHEMICAL SWITCHES

The concept of the redox-switches would be further extended to the electrochemical switches because the redox-active groups are mostly responsible to the electrochemical signals. Basically, the redox-switched crown ethers mentioned above become the latent candidates for this class of crown ethers. The electrochemical switch is sometimes superior to the treatment with redox reagents, because one can keep the system clean as long as the electrochemical reaction is reversible. From a biomimetic viewpoint, to give an electrochemical switch to a crown family is also fascinating because the coupling of the electrochemical switch with transport phenomena mimics many biological events occurring in nerve cells.

Redox-switched crown ethers having the redox-active quinone moiety were synthesized by Misumi's group and Cooper's group,[15,16] and the redox properties in the presence of alkali metal cations were examined in detail. However, these crown compounds were not applied to membrane transport. Presumably, the electrochemical reversibility of the benzoquinone moiety is not high enough to use as ion carriers in the ion transport system. Echeverria et al.[17] developed a new transport cell in which the U-type transport cell is combined with the electrochemical cell (Figure 1) and used an electrochemically more stable anthraquinone derivative (Structure 8) as an ion carrier. The redox-switched carrier Structure 8 is electrochemically reduced to Structure $8^-$ at the IN interface and extracts $Li^+$ as a countercation into the membrane. This process is monitored by the development of a very intense red color, which is quickly distributed exclusively over the organic phase. If electrolysis is stopped, the red color slowly fades and the rate of $Li^+$ transport is slowed down.

Ferrocene-containing crown ethers have been synthesized by several groups.[18-20] It is

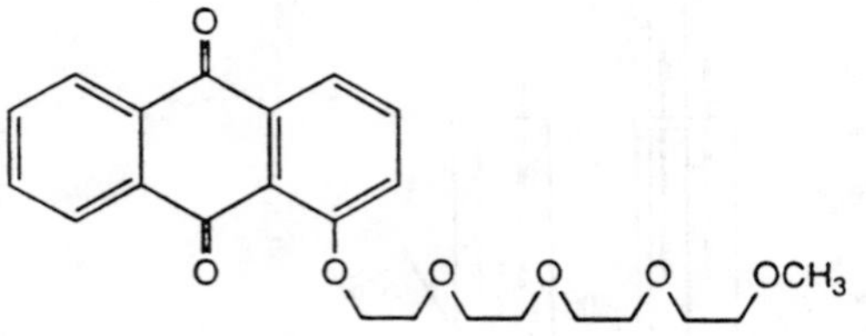

8

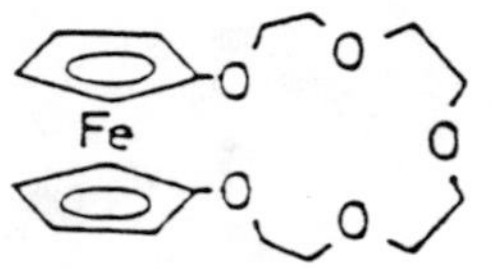

9

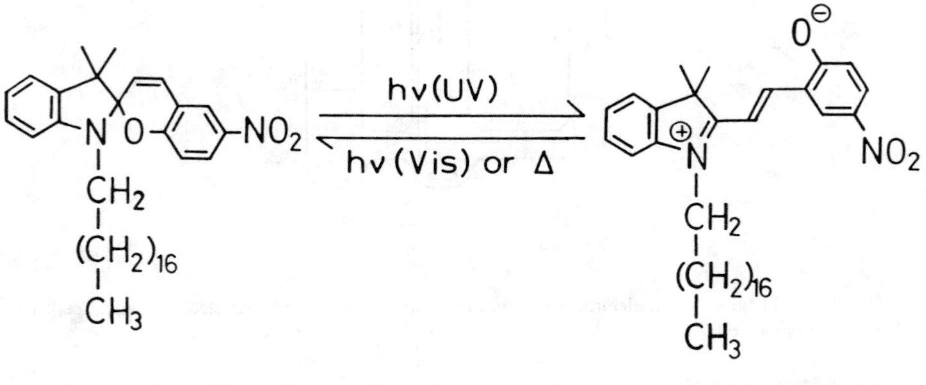

10

$$10^{\pm}$$

known that Structure 9 shows the reversible one-electron oxidation wave at $-0.23$ V (half-wave potential).[21] Thus, the Structure 9-M$^+$ complex would be readily decomposed by oxidation of Structure 9 to Structure 9$^+$. Saji and Kinoshita[22] used Structure 9 as a Na$^+$ carrier across a liquid membrane. When the redox-treatment was carried out for Structure 9 in the membrane, the rate of Na$^+$ transport was speeded up. Probably, the rate of enhancement is caused by the rapid release of Na$^+$ to the OUT aqueous phase.

## IV. PHOTOCHEMICAL SWITCHES

Photoresponsive systems are ubiquitously seen in nature, and light is coupled with the subsequent life processes. In these systems, a photoantenna to capture a photon is skillfully combined with a functional group to mediate some subsequent events. Important is the fact that these events are frequently linked with photoinduced structural changes of photoantennas. This suggests that in an artificial photoresponsive system, chemical substances which exhibit photoinduced structural changes may serve as potential candidates for the photoantennas. The prerequisites for the practical photoantennas would be (a) high quantum yield, (b) high reversibility, and (c) large geometrical or electrostatic change. In the past, photochemical reactions such as (E)-(Z) isomerism of azobenzene, dimerization of anthracene, spiropyran-merocyanine interconversion, etc. have been used as practical photoantennas.

Shimidzu and Yoshikawa[23] used a long-alkyl-group spiropyran compound (Structure 10) as an ion carrier through the liquid membrane (1-octanol). They found that NaCl was concentrated from the dark to the visible light illuminated side across the anisotropic membrane. The mechanism of the Na$^+$ concentration is explained as follows: the opened-carrier (merocyanine form: Structure 10$^+$) on the dark side extracts Na$^+$ into the membrane. The complex formed diffuses from the dark side to the visible light illuminated side with its concentration gradient. On the visible side, the opened-carrier is closed by the visible light and the transported salt is released to the visible light illuminated aqueous phase. The similar system is applied for transport of amino acids across liposomal bilayers.[24]

One can expect that if one of these photoantennas is skillfully combined with a crown ether, many physical and chemical functions of a crown ether family would be controlled by an on-off "light" switch. Shinkai et al.[25] synthesized a variety of photoresponsive crown ethers in which the photofunctional azobenzene is combined intramolecularly with crown ether compounds.

Crown ethers of the type of Structure 11, bearing an intra-annular substituent X, bind metal cations to different extents, depending on the nature of X. When X has no metal-coordination ability, it simply causes steric hindrance and decreases the binding constant. In contrast, when it has a metal-coordination ability (X = OH, COOH, $NH_2$, etc.), it increases the binding constant because of the cooperative action of X and the crown ring. Thus, the molecular design of the compound in which X is switched photochemically between Structures 11 and 12 is very interesting. In case X is the photofunctional azo substituent, one can attain the reversible interconversion between Structures 11 and 12. The azo substituent in Structure (E)-13 simply provides steric hindrance while that in Structure (Z)-13 can coordinate to the metal cation complexed in the photochemically opened crown cavity.[26,27] As expected, the binding constants ($K_s$) of the (Z)-isomers were estimated to be $10^{4.07}$ to $10^{4.81}$ $M^{-1}$ (o-dichlorobenzene − methanol = 5:1 v/v), which are comparable with the $K_s$ for regular crown ethers. The detailed spectroscopic examination showed that the azo group in Structure (Z)-13 coordinates to the complexed metal cation to enhance the $K_s$.[27] In contrast, the (E)-isomers showed no affinity for $Na^+$ and only weakly interacted with $K^+$. This indicates that the (E)-azo group simply provides steric hindrance which rejects metal complexation. Shinkai et al.[27] examined the photoirradiation effect on Structure 13 (n = 1)-mediated $Na^+$ transport across a model organic membrane. Permeation of $Na^+$ was scarcely detected in the dark. In contrast, $Na^+$ was transported under UV light irradiation and the rate was faster than that mediated by dibenzo-18-crown-6. In this system $Na^+$ extraction from the IN aqueous phase to the liquid membrane phase is rate-limiting, and the extraction speed is improved by the photochemical formation of Structure (Z)-13 (n = 1).

In Structure (E)-13 the metal affinity of the crown ring disappears because of steric hindrance. This phenomenon is also realized by competitive complexation of the intramolecular tail with the crown ring. For example, Structure 14 (n = 4, 6, 10) has a crown ring and an ammonium alkyl group attached to the two sides of an azobenzene.[28] These crowns have been designed so that intramolecular "biting" of the ammonium group to the crown can only occur upon photoisomerization to the (Z)-form. Examination of the CPK model suggested that such "tail-biting" can really occur in Structure (Z)-14 (n = 6, 10) but is not the case for Structure (Z)-14 (n = 4) because the spacer $(CH_2)_4$ is too short to effect the "tail-biting". This view was evidenced by the thermal (Z)-to-(E) isomerization: the first-order rate constants were much smaller than those for the free amine counterparts. This suggests that the ammonium tail in Structure (Z)-14 (n = 6, 10) interacts intramolecularly with the crown ether ring. In solvent extraction, the metal affinity for n = 6 and 10 was markedly reduced by UV light irradiation, and in particular the affinity with $K^+$ almost disappeared. This means that the intramolecular ammonium complexation occurs in preference to the intermolecular metal complexation. On the other hand, the metal affinity for n = 4 was less affected by UV light irradiation. In fact, Structure (Z)-14 (n = 4) still extracted $K^+$. In the ion transport system Structure 14 (n = 6, 10) acts as pH-responsive crown ethers as well as photoresponsive crown ethers. With the aid of pH gradient and light they can carry metal cations against the concentration gradient.[28,29] In fact, Structure 14 (n = 6, 10) acted as ion carriers for active transport of $K^+$ from the basic IN aqueous phase to the acidic OUT aqueous phase (Figure 2).[28,29] As expected, active transport of $K^+$ was efficiently speeded up by UV light irradiation.

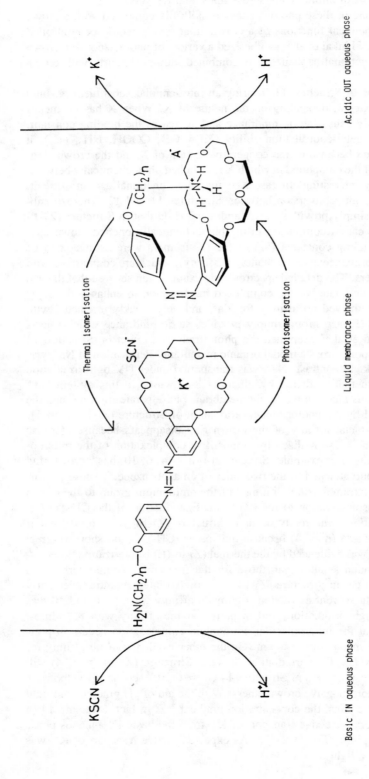

FIGURE 2.   Light-driven K$^+$ transport across a liquid membrane mediated by 14.

11                                    12

(E)-13                                (Z)-13

(E)-14                                (Z)-14

(E)-15                                (E)-16

It has been established that alkali metal cations exactly fitting the size of the crown ether form a 1:1 complex, whereas those having larger cation radii form a 1:2 sandwich complex. It is known that this phenomenon is the main reason to lower the metal selectivity of the crown ether which stems from the principle of the "hole-size selectivity". From a different viewpoint, however, this phenomenon is useful to catch metal cations greater than the crown cavity. This view is clearly substantiated with bis(crown ether)s. For instance, Kimura et al.[30] reported that the maleate diester of monobenzo-15-crown-5 ((Z)-form) extracts K$^+$ from the aqueous phase 14 times more efficiently than the fumarate counterpart ((E)-form). The difference stems from the specific formation of the intramolecular 1:2 sandwich complex with the (Z)-form. If the C=C double bond in these bis(crown ether)s is replaced by the photofunctional azo-linkage, the resultant bis(crown ether)s would exhibit interesting photoresponsive behaviors. That is, the crown rings in the (E)-form would act independently while those in the (Z)-form would act cooperatively to form an intramolecular sandwich

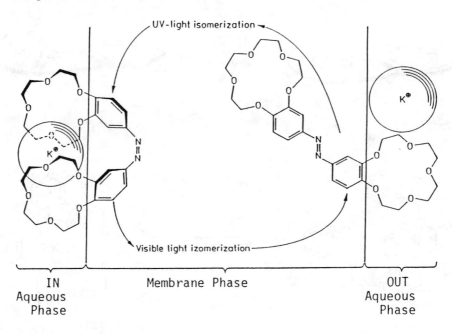

FIGURE 3.    Light-driven K$^+$ transport across a liquid membrane mediated by 15 (n = 2).

complex, and these two states are photochemically interconvertible. The essential idea in this class of photoresponsive crown ethers is that the photoinduced change in the spatial distance between two crown rings should be reflected by the change in the ion-binding ability.

Shinkai et al.[31-34] synthesized a series of azobis(benzocrown ether)s called "butterfly crown ethers", such as Structures 15 and 16. It was found that the content of the (Z)-forms at the photostationary state is remarkably increased with increasing concentration of alkali metal cations which can interact with two crown rings in a 1:2 sandwich manner. Similarly, thermal (Z)-to-(E) isomerization was efficiently suppressed by these metal cations. The findings are both ascribable to the bridge effect of the metal cations with the two crown rings in the (Z)-forms. In two-phase solvent extraction, the (Z)-forms extracted alkali metal cations with large ion radii more efficiently than the corresponding (E)-forms. In particular, the photoirradiation effect on Structure 15 is quite remarkable: for example, Structure (E)-15 (n = 2) extracts Na$^+$ 5.6 times more efficiently than Structure (Z)-15 (n = 2), whereas Structure (Z)-15 (n = 2) extracts K$^+$ 42.5 times more efficiently than Structure (E)-15 (n = 2).[31] Therefore, both the metal affinity and the metal selectivity of azobis(benzocrown ether)s change in response to photoirradiation. The novel finding is applicable to light-driven ion transport across membranes. In K$^+$ transport with Structure 15 (n = 2) across a liquid membrane, it was found that the rate is accelerated by the UV light irradiation which mediates the (E)-to-(Z) isomerization.[31] Thus, the rate enhancement is attributed to the increased extraction speed by Structure (Z)-15 (n = 2) from the IN aqueous phase to the membrane phase. Interestingly, the rate was further enhanced by alternate irradiation by UV and visible (which mediates the (Z)-to-(E) isomerization) light (Figure 3).[34] The finding is rationalized in terms of the increased release speed from the membrane phase to the OUT aqueous phase. Therefore, the rate-limiting step has been switched from the ion extraction to the ion release by alternate light irradiation.

The basic concept described about bis(crown ether)s can be extended to the molecular design of crown ethers with a photoresponsive ionic cap. As mentioned in Chapters 3, 4, and 6, the anionic group acts cooperatively with the crown ring upon extraction of alkali

(E)-17           (Z)-17

(E)-18: $R=COOCH_2CH_2OCH_3$
(E)-19: $R=CO(OCH_2CH_2)_3OCH_3$

and alkaline earth metal cations. In order to exert the cooperativity, the anionic group should be placed exactly on the top of the crown ring. Thus, the possibility arises that the ion affinity is controllable by the change in the spatial position of the intramolecular anionic group. For example, if one can synthesize a crown ether putting on and off an anion-cap in response to photoirradiation, it would exhibit an excellent photoresponsiveness in the ion binding properties. Structure 17 is designed so that a phenolate anion can move to the top of the crown ring upon photoisomerization of the azobenzene segment from the (E)-form to the (Z)-form.[35] The nitro group and the n-butyl group are introduced to lower the $pK_a$ of the phenol group and to enhance the lipophilicity, respectively. Two-phase solvent extraction with 17 showed that the extractability is markedly improved by UV light irradiation which produces the (Z)-form. In particular, $K^+$, $Rb^+$, and $Ca^{2+}$ were extracted efficiently by the (Z)-form. The result clearly indicates that the phenolate anion can act as a cap for the crown ring only when it is isomerized to the (Z)-form. In ion transport across a liquid membrane, the rates of $Na^+$ and $Ca^{2+}$ transport were enhanced by irradiating the liquid membrane phase by UV light, indicating that the anion-capped (Z)-form extracts metal cations into the membrane more efficiently than the (E)-form.

The change in the spatial distance between two ionophoric groups can be also achieved with the photofunctional group other than azobenzene. Irie and Kato[36] designed new photoresponsive ionophores called "molecular tweezers". As illustrated in Structure 18 and Structure 19 the photofunctional group is thioindigo. The (E)-to-(Z) and (Z)-to-(E) isomerization occur reversibly by irradiation of 529 and 488 nm light, respectively. Solvent extraction of metal cations with Structure 18 revealed that the (E)-form has no binding ability to any of the metal cations, whereas $K^+$, $Rb^+$, and $Na^+$ were extracted ($Na^+ < Rb^+ < K^+$) by the photogenerated (Z)-form. In addition, the (Z)-form showed the high binding ability toward heavy metal cations such as $Ag^+$, $Hg^+$, and $Cu^{2+}$. Ion transport experiments showed that repeating the cycles of alternate photoirradiation of 529- and 488-nm light, which causes the (E)-to-(Z) isomerization (529 nm), and the (Z)-to-(E) isomerization (488 nm), results in the transportation of $Ag^+$ across a liquid membrane.[36]

The above examples demonstrate that the rate of carrier-mediated ion transport is controllable by an on-off light switch. The novel phenomenon has been attained by the combination of a photofunctional group with a crown ether which carries metal cations. One may regard, therefore, that pH gradient frequently used for ion-transport by some polyether antibiotics is replaced by light energy in the present light-driven transport systems. Immobilization of photoresponsive crown ethers such as Structures 14, 15, and other photoresponsive crown ethers in thin polymer films is now being studied by Kajiyama et al.[37,38] It has been found that the photoresponsive speed is very fast in these thin membrane systems.

# V. TEMPERATURE SWITCHES

According to Thoman,[40] the rate of carrier-mediated ion transport across a liquid membrane is decreased with increasing temperature. A possible reason for this unusual temperature relationship is related to the fact that the rate-determining event in ionic transport phenomena occurs at the water-membrane interface: that is, when the ion uptake from the first aqueous phase to the membrane phase is a rate-determining step, the overall rate is correlated with the $K_s$ which decreases with increasing temperature. This concept is applicable to active transport of ions driven by the temperature gradient. Oaki and Ishida[41] found that $Na^+$ is transported across an ion-exchange resin membrane from the low temperature side (25°C) to the high temperature side (72°C) in an active transport manner.

It seems to us, however, that the abrupt temperature response would be achieved by using some thermodynamically discontinuous systems. The phase transition phenomena frequently seen in molecular assemblies are the most potential candidates for this purpose. Structure 20 shows a crystal-nematic liquid crystal phase transition at room temperature ($T_{KN}$ 304K) and the liquid crystal phase is very fluid. Thus, this phase transition phenomenon may be useful for the temperature regulation. Shinkai et al.[42] prepared a ternary composite membrane composed of polycarbonate/20/amphiphilic crown ethers (Structures 21 to 24). Above $T_{KN}$ $K^+$ transport mediated by these crown ethers was very fast owing to the high fluidity of Structure 20 forming a continuous phase in the membrane. Thus, one may call this system "an immobilized liquid membrane". Below $T_{KN}$, on the other hand, the rate of $K^+$ transport was efficiently suppressed, indicating that carrier-mediated $K^+$ transport is directly affected by the molecular motion of the liquid crystal phase. Interestingly, when the amphiphilic crown ethers form phase-separated aggregates in the Structure 20 phase (e.g., Structures 22 and 24), $K^+$ transport was suppressed "completely" below $T_{KN}$. The origin of the dramatic rate suppression was explained as follows: in the crystal phase (below $T_{KN}$) diffusion of carriers is no longer possible, but the jump mechanism (ion jumps from one to another carrier fixed in the crystal lattice) is still allowed. This mechanism is actually operative in the system where carriers are dispersed homogeneously because the ion can find the next crown to jump in the neighborhood. When carriers form phase-separated aggregates, this permeation mechanism cannot be operative because the distance between each aggregate is too far to jump. Thus, the ion flux was stopped completely.

The above example suggests the important relation between the phase separation of carriers and the ion transport ability. We noticed that the phase separation would be achieved more easily by fluorocarbon compounds. Structure 25 having a fluorocarbon chain forms phase-separated aggregates in the polycarbonate/Structure 20 composite membrane.[43] As expected, $K^+$ transport through this membrane was "completely" suppressed below $T_{KN}$. In addition, this membrane provided an unexpected transport property characteristic of

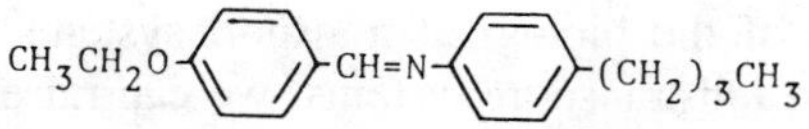

$$CH_3CH_2O-\langle\ \rangle-CH=N-\langle\ \rangle-(CH_2)_3CH_3$$

20

$$R-N \underset{O\ \ \ \ O}{\overset{O\ \ \ \ O}{\quad}} \overset{+}{HN}-CH_2-X$$

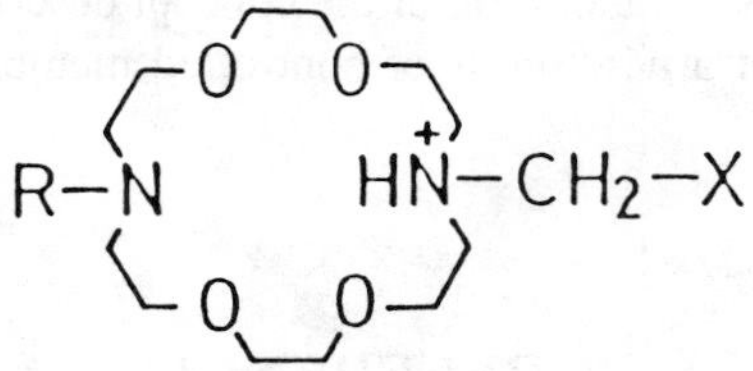

|  | R- |  | -X |
|---|---|---|---|
| 21: | $CH_3(CH_2)nCO-$ | (n = 10, 16) | $-CO_2^-$ |
| 22: | $CH_3(CH_2)_{15}OCH_2$ $CH_3(CH_2)_{15}OCH_2$ $CHOCH_2CO-$ |  | $-CO_2^-$ |
| 23: | $CH_3(CH_2)_{10}CO-$ |  | $-CH_2SO_3^-$ |
| 24: | $CH_3(CH_2)_{15}OCH_2$ $CH_3(CH_2)_{15}OCH_2$ $CHOCH_2CO-$ |  | $-CH_2SO_3^-$ |

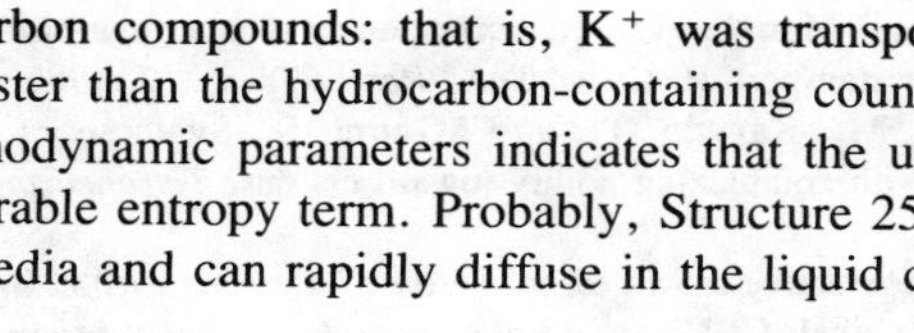

$CF_3(CF_2)_7CH_2CH_2CH_2N$ | $Me(CH_2)_2CO$

25

$Me(CH_2)_{11}N$ | $Me(CH_2)_2C=O$

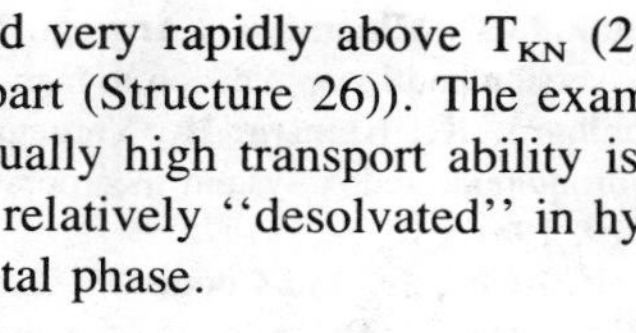

26

fluorocarbon compounds: that is, $K^+$ was transported very rapidly above $T_{KN}$ (21 to 23 times faster than the hydrocarbon-containing counterpart (Structure 26)). The examination of thermodynamic parameters indicates that the unusually high transport ability is due to the favorable entropy term. Probably, Structure 25 is relatively "desolvated" in hydrocarbonic media and can rapidly diffuse in the liquid crystal phase.

## VI. CONCLUSION

In this chapter, we demonstrated how the rate of ion transport can be controlled by the stimuli from the outside world. In the efficient system the responsive function is skillfully combined with the ion-binding site so that the stimuli can be rapidly transduced to the ion-binding site. One may thus consider that the basic idea of the controlled ion-transport systems is very similar to that of the biological transport systems. As future problems to further elaborate the controlled ion-transport systems, we can raise (a) the response speed, (b) the high reversibility, and (c) the control with clean energy sources. From a viewpoint to keep the system clean, light, temperature, magnetic (or electric) field, etc. would be superior to pH, redox reagents, etc. As future possibilities, the transport and the recognition of amino acids, sugars, nucleic acids, etc. will be possibly controlled by the stimuli-responsive systems. We believe that further elaboration of the concept described in this chapter would lead to eventual development of a new series of controlled membrane transport systems.

# REFERENCES

1. **Raymond, K. N. and Carrano, C. J.,** Coordination chemistry and microbial iron transport, *Acc. Chem. Res.,* 12, 183, 1979.
2. **Choy, E. M., Evans, D. F., and Cussler, E. L.,** A selective membrane for transporting sodium ion against its concentration gradient, *J. Am. Chem. Soc.,* 96, 7085, 1984.
3. **Kirch, M. and Lehn, J.-M.,** Selective transport of alkali metal cations through a liquid membrane by macrobicyclic carriers, *Agnew. Chem. Int. Ed. Engl.,* 14, 555, 1975.
4. **Kobuke, Y., Hanji, K., Horiguchi, K., Asada, M., Nakayama, Y., and Furukawa, J.,** Macrocyclic ligands composed of tetrahydrofuran for selective transport of monovalent cations through liquid membrane, *J. Am. Chem. Soc.,* 98, 7414, 1976.
5. **Lamb, J. D., Christensen, J. J., Oscarson, J. L., Nielsen, B. L., Asay, B. W., and Izatt, R. M.,** The relationship between complex stability constants and rates of cation transport through liquid membranes by macrocyclic carriers, *J. Am. Chem. Soc.,* 102, 6820, 1980.
6. **Matsuo, S., Ohki, A., Takagi, M., and Ueno, K.,** Bathocuproine-mediated copper transport through liquid membrane driven by redox potential, *Chem. Lett.,* 1981, 1543.
7. **Ohki, A., Takeda, T., Takagi, M., and Ueno, K.,** Thioether-mediated copper transport through liquid membranes with the aid of redox reaction, *J. Membr. Sci.,* 15, 231, 1983.
8. **Brown, P. R., Izatt, R. M., Christensen, J. J., and Lamb, J. D.,** Transport of $Eu^{2+}$ in a $H_2O$-$CHCl_3$-$H_2O$ liquid membrane system containing the macrocyclic polyether 18-crown-6, *J. Membr. Sci.,* 13, 85, 1983.
9. **Shinbo, T., Kurihara, K., Kobatake, Y., and Kamo, N.,** Active transport of picrate anion through organic liquid membrane, *Nature,* 270, 277, 1977.
10. **Grimaldi, J. J. and Lehn, J.-M.,** Multicarrier transport: coupled transport of electrons and metal cations mediated by an electron carrier and a selective cation carrier, *J. Am. Chem. Soc.,* 101, 1333, 1979.
11. **Shinkai, S., Inuzuka, K., Miyazaki, O., and Manabe, O.,** Redox-switched crown ethers, *J. Org. Chem.,* 49, 3440, 1984.
12. **Shinkai, S., Inuzuka, K., Miyazaki, O., and Manabe, O.,** Cyclic-acyclic interconversion coupled with redox between dithiol and disulfide and its application to membrane transport, *J. Am. Chem. Soc.,* 107, 3950, 1985.
13. **Raban, M., Greenblatt, J., and Kandil, F.,** Switched-on crown ethers, *J. Chem. Soc. Chem. Commun.,* 1983, 1409.
14. **Shinkai, S., Minami, T., Araragi, Y., and Manabe, O.,** Redox-mediated monocrown-biscrown interconversion and its application to membrane transport, *J. Chem. Soc., Perkin Trans.,* 2, 1985, 503.
15. **Sugihara, K., Kamiya, H., Yamaguchi, M., Kaneda, T., and Misumi, S.,** Synthesis of a quinone-hydroquinone redox system incorporated with complexing ability toward cations, *Tetrahedron Lett.,* 22, 1619, 1981.
16. **Wolf, R. E., Jr. and Cooper, S. R.,** Redox-active crown ether, *J. Am. Chem. Soc.,* 106, 4646, 1984.
17. **Echeverria, L., Delgado, M., Gatto, V., Gokel, G. W., and Echegoyen, L.,** Enhanced transport of $Li^+$ through an organic model membrane by an electrochemically reduced anthraquinone podand, *J. Am. Chem. Soc.,* 108, 6825, 1986.
18. **Kasahara, A.,** Syntheses of ferrocenophanes, *Kagaku,* 38, 859, 1983.
19. **Biernat, J. F. and Wilczewski, T.,** Macrocyclic polyfunctional Lewis bases, *Tetrahedron,* 36, 2521, 1980.
20. **Akabori, S., Ohtomi, M., Sato, M., and Ebine, S.,** Syntheses of acyclic polyethers containing ferrocene nucleus and extraction ability, *Bull. Chem. Soc. Jpn.,* 56, 1455, 1983.
21. **Saji, T.,** Electrochemically switched cation binding in pentaoxa[13]ferrocenophane, *Chem. Lett.,* 1986, 275.

22. **Saji, T. and Kinoshita, I.,** Electrochemical ion transport with ferrocene functionalized crown ether, *J. Chem. Soc., Chem. Commun.,* 1986, 716.
23. **Shimidzu, T. and Yoshikawa, M.,** Photo-induced carrier mediated transport of alkali metal salts, *J. Membr. Sci.,* 13, 1, 1983.
24. **Sunamoto, J., Iwamoto, K., Mohri, Y., and Kominato, T.,** Transport of an amino acid across liposomal bilayers as mediated by a photoresponsive carrier, *J. Am. Chem. Soc.,* 104, 5502, 1982.
25. **Shinkai, S. and Manabe, O.,** Photocontrol of ion extraction and ion transport by photofunctional crown ethers, *Top. Curr. Chem.,* 121, 67, 1984.
26. **Shinkai, S., Miyazaki, K., and Manabe, O.,** A photochemically "switched-on" crown ether containing an intraannular 4-methoxyphenylazo substituent, *Angew. Chem. Int. Ed. Engl.,* 24, 866, 1985.
27. **Shinkai, S., Miyazaki, K., and Manabe, O.,** Photochemically switched-on crown ethers containing an intraannular azo substituent and their application to membrane transport, *J. Chem. Soc., Perkin Trans.,* 1, 1987, 499.
28. **Shinkai, S., Ishihara, M., Ueda, K., and Manabe, O.,** Photoregulated crown-metal complexation by competitive intramolecular tail(ammonium)-biting, *J. Chem. Soc., Perkin Trans.,* 2, 1985, 511.
29. **Shinkai, S., Ishihara, M., Ueda, K., and Manabe, O.,** On-off-switched crown-metal complexation by photoinduced intramolecular tail(ammonium)-biting, *J. Inclusion Phenom.,* 2, 111, 1984.
30. **Kimura, K., Tamura, H., Tsuchida, T., and Shono, T.,** New bis(crown ether)s derived from stereo-isomers of dicarboxylic acid, *Chem. Lett.,* 1979, 611.
31. **Shinkai, S., Nakaji, T., Ogawa, T., Shigematsu, K., and Manabe, O.,** Photocontrol of ion extraction and ion transport by a bis(crown ether) with a butterfly-like motion, *J. Am. Chem. Soc.,* 103, 111, 1981.
32. **Shinkai, S., Shigematsu, K., Kusano, Y., and Manabe, O.,** Photocontrol of ion extraction and ion transport by several photofunctional bis(crown ethers), *J. Chem. Soc., Perkin Trans.,* 1, 1981, 3279.
33. **Shinkai, S., Ogawa, T., Kusano, Y., Manabe, O., Kikukawa, K., Goto, T., and Matsuda, T.,** Influence of alkali metal cations on photoisomerization and thermal isomerization of azobis(benzocrown ether)s, *J. Am. Chem. Soc.,* 104, 1960, 1982.
34. **Shinkai, S., Shigematsu, K., Sato, M., and Manabe, O.,** Ion transport mediated by photoinduced cis-trans interconversion of azobis(benzocrown ether)s, *J. Chem. Soc., Perkin Trans.,* 1, 1982, 2735.
35. **Shinkai, S., Minami, T., Kusano, Y., and Manabe, O.,** Light-driven Ion transport by crown ethers with a photoresponsive anionic cap, *J. Am. Chem. Soc.,* 104, 1967, 1982.
36. **Irie, M. and Kato, M.,** Photoresponsive molecular tweezers, *J. Am. Chem. Soc.,* 107, 1024, 1985.
37. **Kumano, A., Niwa, O., Kajiyama, T., Takayanagi, M., Kano, K., and Shinkai, S.,** Photoinduced permeation through ternary composite membrane composed of polymer/liquid crystal/azobenzene-bridged crown ether, *Chem. Lett.,* 1983, 1327.
38. **Kajiyama, T., Kikuchi, H., Terada, I., Katayose, M., Takahara, A., and Shinkai, S.,** Permeation properties of polymer (liquid crystal) composite membrane, *Curr. Top. Polym. Sci.,* 2, 319, 1987.
39. **Anzai, J., Ueno, A., Sasaki, H., Shimokawa, H., and Osa, T.,** Photocontrolled permeation of alkali cations through poly(vinyl chloride)/crown ether membrane, *Makromol. Chem., Rapid Commun.,* 4, 731, 1983.
40. **Thomon, C. J.,** Effect of temperature on the transport capabilities of some common ionophores, *J. Am. Chem. Soc.,* 107, 1437, 1985.
41. **Oaki, H. and Ishida, M.,** Active transport of sodium chloride by a thermal regeneration ion-exchange resin membrane when temperature gradient is applied as energy source, *Maku (Membr.),* 5, 371, 1980.
42. **Shinkai, S., Nakamura, S., Ohara, K., Tachiki, S., Manabe, O., and Kajiyama, T.,** "Complete" thermocontrol of ion permeation through ternary composite membranes composed of polymer/liquid crystal/ amphiphilic crown ether, *Macromolecules,* 20, 21, 1987.
43. **Shinkai, S., Torigoe, K., Manabe, O., and Kajiyama, T.,** Temperature regulation of crown-mediated ion transport through polymer/liquid crystal composite membranes, *J. Am. Chem. Soc.,* 109, 4458, 1987.

Chapter 7.4

# PERMEATION CONTROL THROUGH LIPID BILAYER-IMMOBILIZED FILMS

**Yoshio Okahata**

## TABLE OF CONTENTS

# I. INTRODUCTION

Cell membranes are mainly composed from a lipid bilayer matrix of phospholipids, proteins incorporated either on the surface or in the interior, and polysaccharides adsorbed mainly on the surface. Complexities of natural membranes have necessitated the use of simple models for investigations at the molecular level. Liposomes, prepared from naturally occurring phospholipids, have been widely used as a simple closed membrane analog.[1-3] A decade ago, Kunitake and Okahata,[4,5] and then other investigators,[6-13] have reported that lipid bilayer vesicles can be formed from various kinds of synthetic single-chain, double-chain, and triple-chain amphiphiles having cationic, anionic, and nonionic polar head groups.

Liposomes, as well as synthetic bilayer vesicles, can entrap water-soluble substances in the inner aqueous phase, retain them for extended periods, and release them by a function of the phase transition of bilayers or other outside effects. Although they are stable under some static conditions, vesicles are liable to fuse with each other, and their bilayer walls may be too weak and frangible under the dynamic changes of outside effects such as temperatures near the phase transition, ambient pH, ionic strength, and osmotic pressures. These are serious drawbacks of lipid bilayer vesicles for kinetic studies of permeability.

The strength of the bilayer wall and the prevention of the vesicle fusion have been recently improved by some methods of polymerizations of lipid bilayers[14-19] or by covering vesicles with polymers.[20,21] To overcome the weakness of the lipid wall and a small inner aqueous phase of liposomes, we have developed a lipid bilayer-corked capsule membrane.[22-28] The capsule is formed by a physically strong porous nylon membrane which is corked with multiple lipid bilayers. The corking lipids act as a permeation valve responding to various outside effects.

Recently, Kajiyama, Kunitake, and Okahata and their co-workers reported the preparation of some kinds of self-standing multibilayer-immobilized films: (1) casting of a mixed solution of polymers and amphiphiles and then evaporating solvent slowly,[29-34] (2) corking amphiphiles into a porous filter membrane by dipping the filter into amphiphile solution and drying,[35-37] (3) polymerization of an aqueous cast film of amphiphile multibilayers,[38,39] and (4) preparing a polyion complex from cationic bilayer-forming amphiphiles and polyanions, casting a chloroform solution of the complex on a substrate, and drying.[40-42] Among these methods, the (4) polyion complex method gives the stable multibilayer-immobilized film suitable for permeation measurements in aqueous solution over a wide temperature range, because each amphiphile is immobilized with polymer chain at the head group by the ion interaction and still retains the mobility of dialkyl chains in the film.

In this article, we would like to review signal-receptive permeability control by using polyion complex type multibilayer-immobilized films (see Figure 1). Permeation of ions and water-soluble probes through the film could be controlled reversibly responding to various stimuli from outside such as temperatures,[44-47] ambient pH changes,[45] electric fields,[46,47] and electrochemical reactions.[43,44,48-50] These reversible permeability controls could be observed only by using polyion complex films because of their physical stability and well-oriented fluid multibilayer structures in the film.

# II. PREPARATION OF MULTIBILAYER-IMMOBILIZED FILMS

Preparations of the polyion complex type bilayer-immobilized film from simple dialkylammonium amphiphiles and poly(styrene sulfonate) are shown below as a typical example.

An aqueous dispersion (20 ml) of dioctadecyldimethylammonium bromide ($2C_{18}$ $N^{+}2C_{1}Br^{-}$, 0.5 mmol) and an aqueous solution (20 ml) of sodium poly(styrene sulfonate) ($PSS^{-} Na^{+}$, 0.5 unit mmol) were mixed at 70°C. White precipitates were washed, dried, and purified by reprecipitation in chloroform with methanol. The obtained powder (recovery

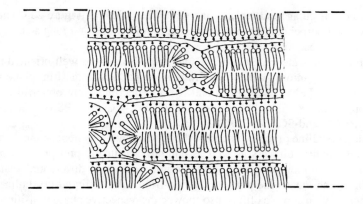

FIGURE 1. A schematic illustration of intersectional view of polyion complex type multibilayer-immobilized cast film, in which each bilayer-forming amphiphile is immobilized with polymer chain at the head group by the ion interaction.

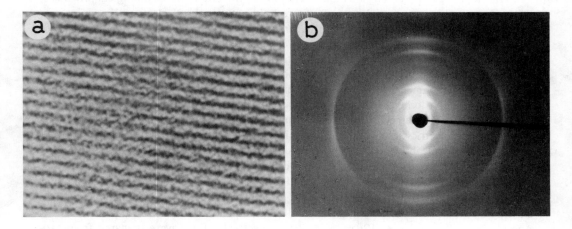

FIGURE 2. Transmission electron micrograph stained negatively with uranyl acetate (a) and X-ray diffraction pattern (b) of the intersection of the $2C_{18}N^+2C_1/PSS^-$ film.

90%) was confirmed from elemental analyses of C, N, S, and halogen that counter bromide ions exchanged to $PSS^-$, and inorganic salts ($Na^+$ and $Br^-$) were not detected within analytical errors. The polyion complex powder ($2C_{18}N^+2C_1/PSS^-$) was dissolved in chloroform and cast on a glass plate. The self-standing transparent film obtained was physically strong and water-insoluble. For permeation experiments, the film was cast on a polyester or platinum minigrid sheet (100 or 200 mesh, 70 μm thick) attached at the bottom of a polyethylene tube by a dipping process in the chloroform solution (see Figure 3). The film thickness was estimated to be 100 μm from the scanning electron micrograph (SEM) observation.

Figure 2a is a transmission electron micrograph (TEM) of the intersectional view of the $2C_{18}N^+2C_1/PSS^-$ film stained negatively, and shows the distinct multibilayer structure whose mean thickness of one white-line is estimated to be *ca.* 3.8 nm. X-Ray diffraction patterns of the $2C_{18}N^+2C_1/PSS^-$ film are shown in Figure 2b, in which the incident beam was exposed parallel to the film plane (edge view). The series of strong reflection arcs with 3.6 nm spacing parallel to the film plane were clearly observed, which consists with both a bimolecular length of $2C_{18}N^+2C_1$ amphiphiles (3.6 to 4.0 nm, estimated from Corey-

Pauling-Kolton molecular models) and one white-line width of Figure 2a. When the X-ray beam was exposed perpendicular to the film plane, reflections of a long spacing of a bilayer thickness were not observed.

These findings indicate that $2C_{18}N^+2C_1$ amphiphiles exist as well-oriented multibilayers complexing with polyanions, which pile growing up parallel to the film plane as illustrated in Figure 1. Similar TEM observations and X-ray reflections were obtained in the cases of other polyion complex films: $C_n$-Azo-$C_m$-$N^+$/PSS$^-$, $2C_{16}V^{2+}$/PSS$^-$, $2C_{16}N^+$Fc/PSS$^-$, FcC$_{11}$C$_{18}$N$^+$/PSS$^-$, and $3C_{12}N^+C_1$/PSS$^-$.

The liquid crystalline property is one of the fundamental physicochemical characteristics of synthetic ammonium bilayers as well as naturally occurring phospholipid bilayers. The $2C_{18}N^+2C_1$/PSS$^-$ film shows a sharp endothermic peak from differential scanning calorimetry (DSC), which means the phase transition from solid to liquid crystalline state of the bilayers. Other polyion complex films also showed the respective phase transition temperature ($T_c$) by DSC measurements.

## III. ELECTRIC FIELD SENSITIVE PERMEATION[46,47]

Transmembrane potentials are believed to play an important role in biological processes. When the membrane potential exceeds a threshold value, a local electrical breakdown (a formation of transient pores) seems to occur in lipid bilayers, allowing rapid passage of large substances and particles (up to the size of genes) which cannot normally permeate through the membrane.[51,52] Extensive pulsation experiments of transmembrane potentials have been concluded on such convenient models as suspension of liposomes[53] and planner lipid bilayer membranes.[54,55] However, in contrast to biological membranes, in the case of these artificial lipid bilayers it has been impossible to induce stable transient pores which have not been accompanied by mechanical rupture of the membrane, for any long duration (>1 s) by the electric field. When the physically stable multibilayer film was employed, the reversible permeability control by an electric field can be expected.

For this purpose, we prepared two kinds of multibilayer-immobilized films from simple dialkylammonium amphiphiles ($2C_{18}N^+2C_1$/PSS$^-$) and liquid crystalline type amphiphiles containing an azobenzene chromophore ($C_n$-Azo-$C_m$-$N^+$/PSS$^-$) and studied permeation changes by an electric potential applied parallel or perpendicular to the film plane. The $C_n$-Azo-$C_m$-$N^+$/PSS$^-$ (n = 8, 12; m = 4, 10, respectively) films were prepared from azobenzene-containing amphiphiles and poly(styrene sulfonate) according to the described method. The azobenzene amphiphiles in the film were confirmed to have multiple bilayer structures parallel to the film plane, as well as $2C_{18}N^+2C_1$ amphiphiles.

The permeation of the water-soluble, nonionic fluorescent probe $NQ_2(1 \times 10^{-3} M)$ through the film cast on a polyester mesh was followed fluorophotometrically at $\lambda_{max}$ 340 nm (excitation at 280 nm) as shown in Figure 3. Permeation rates were obtained both in the absence and presence of an electric field, which was applied either perpendicular (method A) or parallel (method B) to the film plane. A nonionic water-soluble fluorescent probe, $NQ_2$, was chosen as a permeant to avoid the effect of the electrostatic interaction with the electrodes and the polyionic bilayer-films. The relative permeation rate, P (cm$^2$ s$^{-1}$), was obtained from the following equation[40-50]:

$$P = \frac{Jd}{C_0S} \tag{1}$$

where d and S are a thickness (100 μm) and a area (57 mm$^2$) of the film, respectively. J and $C_0$ are a flux of permeation and an intial concentration (1.0 × 10$^{-3}$ M) of the $NQ_2$ probe in the upper cell, respectively.

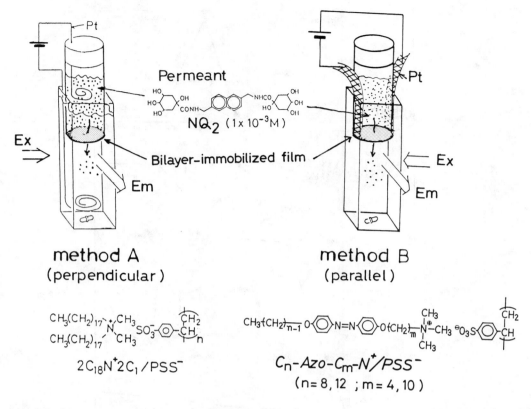

FIGURE 3.   Permeation control through bilayer-immobilized films ($2C_{18}N^+2C_1/PSS^-$ and $C_n$-Azo-$C_m$-$N^+$/$PSS^-$) with an electric field perpendicular (method A) or parallel (method B) to the film.

Typical permeation changes through the dialkyl type $2C_{18}N^+2C_1/PSS^-$ film are shown in Figure 4. The permeation of $NQ_2$ probes was very slow in the absence of an electric field ($P = 1.5 \times 10^{-9}$ cm$^2$ s$^{-1}$), but with an applied voltage of 3 V between two platinum electrodes across the membrane (method A) at 55°C (above $T_c$ of bilayers), permeability was immediately enhanced by a factor of 10 ($P = 1.7 \times 10^{-8}$ cm$^2$ s$^{-1}$) and reverted to the original slow rate when the potential was turned off even after a long duration (1 to 20 min). This permeability change could be reproduced repeatedly without damaging the bilayer-immobilized film at 55°C. This permeability enhancement at 55°C was observed only with a voltage of $>1.7$ V between the two electrodes (corresponding to a voltage of 0.5 V across the membrane with a thickness of 100 μm; 50 V/cm), and increased proportionally with increasing electric field in the range 1.7 to 4.0 V.[46]

The permeation of $NQ_2$ probe at 25°C was not affected by the electric field in the range 0 to 6 V in contrast to the effects at 55°C. Experiments at various temperatures showed that the electro-sensitive permeation occurred only in the fluid liquid crystalline state of the bilayer-film above $T_c$, but not in the solid state below $T_c$. The electro-sensitive permeation at 55°C (above $T_c$), but not at 25°C (below $T_c$), can thus be due to the phase transition of the bilayers in the $2C_{18}N^+2C_1/PSS^-$ film.

When the potential of 3 V was applied parallel to the film (method B), the permeability of $2C_{18}N^+2C_1/PSS^-$ film was hardly increased (see Figure 4).

In the case of the film prepared from azobenzene-containing amphiphiles ($C_{12}$-Azo-$C_4$-$N^+$/$PSS^-$), the permeability enhanced (5 to 10 times) with an applied voltage of 4 V (250 μA) between two electrodes parallel to the membrane (method B) (Figure 5).[47] The permeability reverted to the original slow rate when the potential was turned off even after the

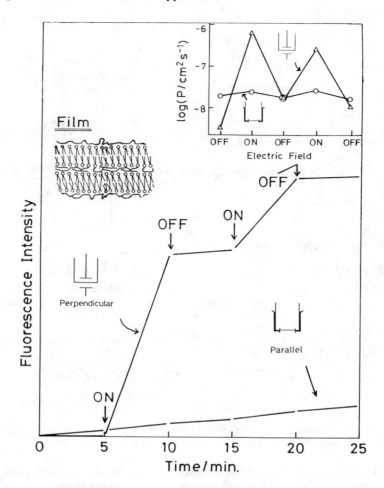

FIGURE 4.    Typical permeation changes through the $2C_{18}N^+2C_1/PSS^-$ film with perpendicular or parallel electric fields at 55°C (above $T_c$ of bilayers).

long charging time (5 to 10 min.), independent of the alkyl chain length of $C_n$-Azo-$C_m$-$N^+$ amphiphiles. This permeability change could be observed at various temperatures in the range of 10 to 60°C independent of the phase transition of bilayers, and could be reproduced repeatedly without damaging $C_n$-Azo-$C_m$-$N^+$/$PSS^-$ films. The permeability enhancement was observed only with a voltage above 1.5 V (50 μA) between two electrodes and increased proportionally with increasing the potential in the range of 1.5 to 8.0 V (50 to 500 μA).

When the potential of 4 V (250 μA) was applied across the membrane (method A), the permeability of $C_n$-Azo-$C_m$-$N^+$/$PSS^-$ films was only slightly increased by a factor of 1.3 to 1.8 (see Figure 5).

These rate enhancements, depending on the direction of electric fields and lipid structures, are explained as follows. Under the electric field across the membrane, the dialkyl-type $2C_{18}N^+2C_1/PSS^-$ film acts as a capacitor because of its poor conductivity, and the electric transmembrane potential can produce transient pores in the fluid bilayers by electrically induced "breakdown" which results in the permeability enhancement. In contrast, the $C_n$-Azo-$C_m$-$N^+$/$PSS^-$ films contain highly conductive chromophores in the bilayers, so that a high transmembrane potential cannot form and no "breakdown" occurs.

When the potential was applied parallel to the $C_n$-Azo-$C_m$-$N^+$/$PSS^-$ film, the orientation of the azobenzene chromophore in the bilayers presumably changes, and the permeability

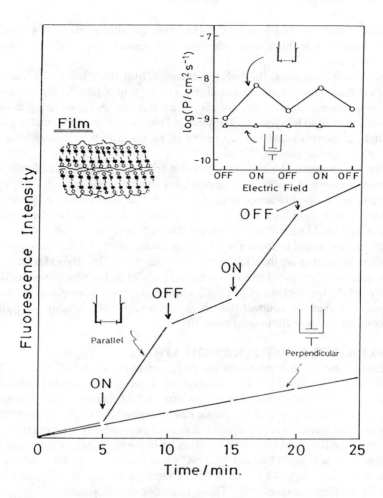

FIGURE 5. Typical permeation changes through $C_{12}$-Azo-$C_4$-N$^+$/PSS$^-$ film with perpendicular or parallel electric fields at 40°C.

of the resulting disordered bilayers increases. The permeability of $2C_{18}N^+2C_1$/PSS$^-$ films did not change when the electric field was applied parallel to the membrane (method B) because the poorly conductive "dialkyl" bilayers are not oriented by the electric field. Thus, the permeability of the bilayer-immobilized film containing the conductive, liquid crystalline chromophores changes when the electric field is applied parallel to the membrane; on the contrary, the permeability of the film from the poorly conductive, dialkyl bilayers changes when the electric field is applied perpendicular to the film.

The fluid bilayer structure is very important for the electrically controlled change in permeability, because the change is not observed when the rigid, polymeric liquid crystals such as poly(benzyl glutamate) and usual polymer films such as polystyrene and acetyl cellulose are employed. The $2C_{18}N^+2C_1$ and $C_n$-Azo-$C_m$-N$^+$ bilayers are immobilized at the hydrophilic head group in contact with polyanions, but still have the fluid bilayer properties.

## IV. BILAYER-INTERCALATED CLAY FILMS[45]

In the previous section, we described the polyion complex film prepared from cationic amphiphiles and an anionic linear poly(styrene sulfonate). The polyion complex film prepared from linear polyanions is stable and reused under mild experimental conditions (pH 4 to 9, ionic strength $<0.05 M$, and 5 to 60°C). However, the film tends to swell and destroy in harsh conditions such as high ionic strength, high temperature, and very basic or acidic conditions for a long time.

In this section, we show the polyion complex film from layer-type montmorillonite anions and cationic bilayer-forming amphiphiles in order to obtain the physically strong film and study as a gate membrane behavior for permeation changes responding to temperatures (phase transition of lipid bilayers) and ambient pH changes. It is known that mica-type layer-silicates such as montmorillonite are capable of exchanging their cations ($Na^+$, $K^+$, and $Ca^{2+}$) for other organic cations.[56-59]

The film was prepared by mixing equimolar amounts of aqueous colloidal solutions of clay montmorillonite and ammonium amphiphiles similar to the polyion complex film with linear polyanions. X-ray diffraction pattern of $2C_{18}N^+2C_1$-intercalated clay film showed the series of strong reflection arcs with 4.38 nm corresponding to basal plane spacing of the amphiphile-intercalated layer-silicate. From the obtained basal spacing of 4.38 nm, $2C_{18}N^+2C_1$ amphiphiles are estimated to form a single bilayer structure (3.85 nm thick with a tilt angle of 55°) in the interlayer of the clay (0.98 nm thick[55-59]). This indicates that $2C_{18}N^+2C_1$ bilayer-intercalated clay layers piles growing up parallel to the film plane, as illustrated in Figure 6. When the triple-chain $3C_{12}N^+3C_1$ amphiphiles were employed, the similar bilayer structure was observed. In contrast, the single-chain $C_{16}N^+3C_1$ amphiphile forms a mono-layer structure in the clay film (see Figure 6).

## A. TEMPERATURE SENSITIVE PERMEATION

Permeation rates of a nonionic probe $NQ_1$ across the $C_{16}N^+3C_1$/clay$^-$, $2C_{16}N^+2C_1$/clay$^-$, and $3C_{12}N^+3C_1$/clay$^-$ films were obtained at various temperatures below and above their $T_c$ to study the effect of the phase transition on $NQ_1$ permeation. Arrhenius plots are shown in Figure 7. In the case of the single chain amphiphile, $C_{16}N^+3C_1$-intercalated film, the Arrhenius plot gave a simple straight line, and permeability was relatively high. The $C_{16}N^+3C_1$ amphiphiles form a monolayer structure in the interlayer, and the phase transition of lipid structures was not observed by DSC measurements. On the contrary, Arrhenius plots gave inflections near 43°C and 35°C in the case of $2C_{16}N^+2C_1$- and $3C_{12}N^+3C_1$-intercalated clay films, respectively. These inflections correspond to $T_c$ of the respective bilayers obtained by DSC measurements (shown by arrows in the figure). The drastic increases in $NQ_1$ permeation at $T_c$ are associated with the phase transition of the intercalated bilayers from a well-oriented solid state to a disordered liquid crystalline state.

## B. pH SENSITIVE PERMEATION

The clay film intercalated with dissociative bilayer-forming amphiphiles having primary amino head groups, $2C_{16}$-$NH_3^+$, can be expected to change the permeability responding to ambient pH changes. The film was prepared in three different ways as shown schematically in Figure 8: (a) an equimolar amount of aqueous solutions of $2C_{16}$-$NH_3^+$ and $2C_{16}N^+2C_1$ amphiphiles were mixed with a montmorillonite solution, and the chloroform solution of the precipitate was cast as a film [($2C_{16}$-$NH_3^+ + 2C_{16}N^+2C_1$)/clay$^-$] in which each amphi-phile was mixed in a clay layer, (b) equimolar amounts of chloroform solutions of $2C_{16}$-$NH_3^+$/clay$^-$ and of $2C_{16}N^+2C_1$/clay$^-$ were mixed and cast as a film [($2C_{16}$-$NH_3^+$/clay$^-$) + ($2C_{16}N^+2C_1$/clay$^-$)] in which each amphiphile was not mixed in a clay layer, and (c) the film was prepared from only dissociative $2C_{16}$-$NH_3^+$ amphiphiles ($2C_{16}$-$NH_3^+$/clay$^-$ film).

When the clay film was prepared by the method (a) [($2C_{16}$-$NH_3^+ + 2C_{16}N^+2C_1$)/clay], typical time courses of permeation of $NQ_1$ probes are shown in Figure 9. pH of the lower aqueous solution was changed by the addition of aliquots of aq. NaOH and aq. HCl, repeatedly. The permeability was slow and not affected by changing the pH from 7 to 11

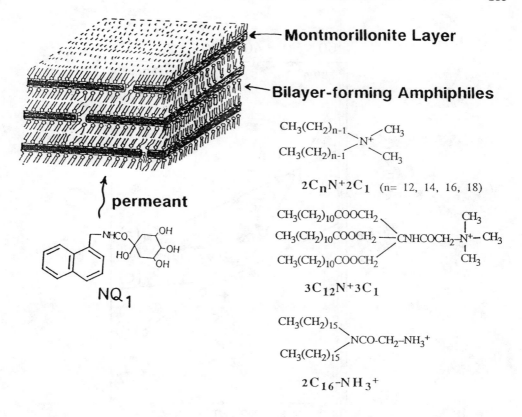

**Montmorillonite Layer**

**Bilayer-forming Amphiphiles**

permeant

$NQ_1$

$2C_nN^+2C_1$ (n= 12, 14, 16, 18)

$3C_{12}N^+3C_1$

$2C_{16}$-$NH_3^+$

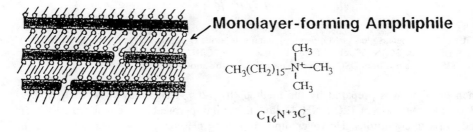

**Monolayer-forming Amphiphile**

$C_{16}N^+3C_1$

FIGURE 6.   Illustration of lipid bilayer-intercalated clay films. In the case of single-chain amphiphile, $C_{16}N^+3C_1$ forms a monolayer in the interlayer instead of a bilayer structure.

at the first stage. Upon changing repeatedly pH between 7 and 11 several times, the permeability at pH 11 became large compared with that at pH 7. The permeability changes between pH 7 and 11 became reversible after 4 to 5 repeats of pH changes without damaging the film at least 20 times.

The dissociation of head groups of $2C_{16}$-$NH_3^+$ amphiphiles in the film seems to play an important role for the observed pH-sensitive permeations. pH-rate profiles of the probe permeation were studied at 20, 40, and 60°C and shown in Figure 10. The permeability increased greatly above pH 9 independent of temperatures, suggesting that the neutralization of $-NH_3^+$ cationic head groups to $-NH_2$ groups occur both in fluid and solid states of bilayers ($T_c$ = 52°C). Thus, $2C_{16}$-$NH_3^+$ bilayers in the ($2C_{16}$-$NH_3^+$ + $2C_{16}N^+2C_1$)/clay$^-$ film provide a high barrier to the probe permeation in the cationic bilayer form below pH 9, but not in the neutral form of $2C_{16}$-$NH_2$ above pH 9.

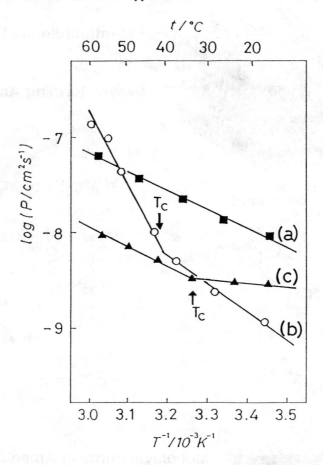

FIGURE 7.    Arrhenius plots of $NQ_1$ permeation through (a) the $C_{16}N^+3C_1/clay^-$, (b) the $2C_{16}N^+2C_1/clay^-$, and (c) the $3C_{12}N^+3C_1/clay^-$ films. $T_c$ values of the $2C_{16}N^+2C_1/clay^-$ (43°C) and $3C_{12}N^+3C_1/clay^-$ (35°C) obtained from DSC measurements are shown by arrows, respectively.

When the film was prepared by the method (b) $[(2C_{16}-NH_3^+/clay^-) + (2C_{16}N^+2C_1/clay^-)]$ or the method (c) $(2C_{16}-NH_3^+/clay^-)$, similar permeability enhancements were observed in basic conditions. However, upon repeating pH changes between 7 and 11, the increased permeability did not revert to the original slow rate at pH 7, and the film became broken in the basic condition. Thus, in the $2C_{16}-NH_3^+/clay^-$ film prepared from only dissociative amphiphiles [method (c)], the neutralized $2C_{16}-NH_2$ amphiphiles may release easily from the interlayer of the clay film because of the disappearance of the electrostatic interaction between amphiphiles and the anionic silicate-layer. In the case of $[(2C_{16}-NH_3^+/clay^-) + (2C_{16}N^+2C_1/clay^-)]$ film prepared from method (b), both amphiphiles are not mixed with each other in a clay layer, and the neutralized $2C_{16}-NH_2$-intercalated layers may be broken at pH 11. On the contrary, the $(2C_{16}-NH_3^+ + 2C_{16}N^+2C_1)/clay^-$ film prepared from method (a) was physically stable even after repeating pH changes between 7 and 11 for many times, because the $2C_{16}-NH_3^+$ and $2C_{16}N^+2C_1$ amphiphiles are completely mixed with each other in a clay layer and the interaction between the cationic $2C_{16}N^+2C_1$ amphiphiles and the anionic silicate-layer keeps the film structure even when the dissociative amphiphiles are neutralized.

When the polyion complex film was prepared from a linear poly(styrene sulfonate) with the amphiphile mixture of $2C_{16}-NH_3^+$ and $2C_{16}N^+2C_1$ according to method (a) and was employed as a pH-sensitive permeation film, the film was easily broken at the alkaline

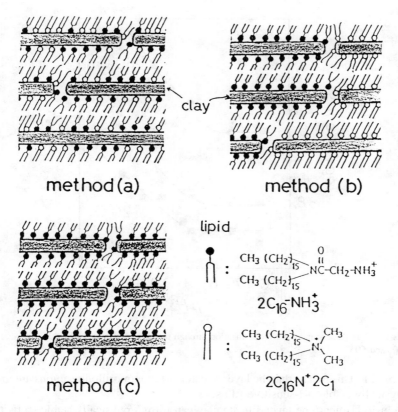

FIGURE 8. Three different types of pH-sensitive clay films intercalated with dissociative $2C_{16}$-$NH_3^+$ bilayers. Method (a): each amphiphile of $2C_{16}$-$NH_3^+$ and $2C_{16}N^+2C_1$ is completely mixed in a silicate layer, method (b): each amphiphile is separated in each clay layer, and method (c): the film is prepared from only $2C_{16}$-$NH_3^+$ amphiphiles.

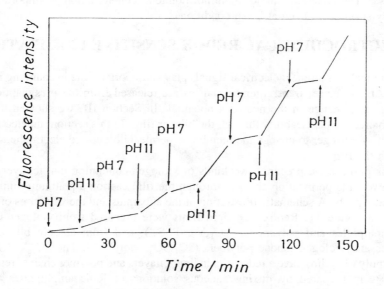

FIGURE 9. Typical time courses of $NQ_1$ permeation through the $(2C_{16}$-$NH_3^+$ + $2C_{16}N^+2C_1)$/clay$^-$ film (method (a)) responding to pH changes between 7 and 11 at 60°C (above $T_c$).

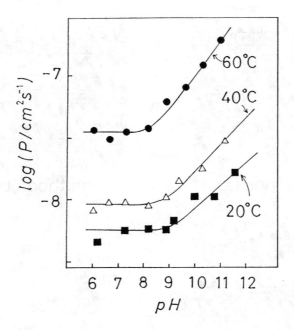

FIGURE 10.    pH-Dependence of $NQ_1$ permeation through the $(2C_{16}-NH_3^+ + 2C_{16}N^+2C_1)/clay^-$ film at 20, 40, and 60°C.

solution of pH 11. This means that the layer structure of montmorillonite anions are important to form physically stable pH-sensitive films.

In conclusion, mica-type layer-silicate clay can form a physically stable, self-standing, bilayer-intercalated film by simply mixing cationic double- or triple-chain amphiphiles. Permeations through the clay film can be reversibly changed responding to the phase transition temperature of lipid bilayers and ambient pH changes. The permeability could be also controlled by the application of an electric field perpendicular to the film as well as $2C_{18}N^+2C_1/$ PSS$^-$ films.[45] The layer structure of montmorillonite as counter anions is important to form a physically strong film even at harsh conditions.

# V. ELECTROCHEMICAL REDOX SENSITIVE PERMEATION[48,49]

In biological membranes, electrical signals play an important role in changing membrane transport. At a synapse, transmitter substances are released from the presynaptic terminal in response to a change in the neuron's potential. In Section III, we showed the electric field can change the permeability through the bilayer film. In this section, an electric current can also act as a trigger to change the membrane permeability due to electrochemical redox reactions in the film.

For this purpose we prepared two kinds of viologen-containing multibilayer films and achieved reversible permeation changes through the film cast on a platinum minigrid sheet by electrical signals. A schematic illustration of the apparatus and the structures of two films are shown in Figure 11. Redox-type lipid films were prepared from viologen-containing dialkyl amphiphiles and polyanions ($2C_{16}V^{2+}$/PSS$^-$), and from anionic bilayer-forming amphiphiles and viologen ionene polymers ($2C_{12}SO_3^-$/polyV$^{2+}$). The $2C_{16}V^{2+}$/PSS$^-$ and $2C_{12}SO_3^-$/polyV$^{2+}$ films have a redox site in lipid bilayers and polymer chains, respectively. The films were prepared by mixing aqueous solutions of ionic amphiphiles and linear polyelectrolytes similar to other polyion complex films. The film was cast on a platinum minigrid sheet (100 mesh, 70 μm thick) attached at the bottom of a polyethylene tube.

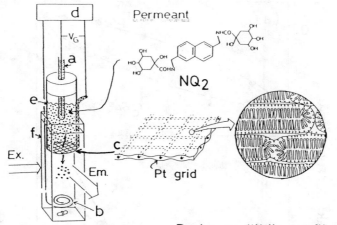

Redox multibilayer film

$2C_{16}V^{2+}/PSS^-$  $2C_{16}V^{+\bullet}/PSS^-$
$(T_C = 24\,°C)$  $(T_C = 38\,°C)$

**FIGURE 11.** Experimental setup for permeation of fluorescent probes $NQ_2$ through redox-type multibilayer films cast on a Pt minigrid sheet. (a) Reference electrode, Ag/AgCl in saturated KCl ($-0.05$ V vs. SCE); (b) Pt wire for counter electrode; (c) Pt minigrid sheet (100 mesh, 70 μm thick, 28 mm² area) embedded in redox-type bilayer films; (d) Potentiostat; (e) Polyethylene tube (diameter, 7 mm); (f) 1-cm quartz cell.

## A. CYCLIC VOLTAMMETRY

Cyclic voltammetry of a $2C_{16}V^{2+}/PSS^-$ film cast on a Pt wire in $N_2$-purged 0.1 $M$ $NaClO_4$ aqueous solution showed the reversible one-electron reduction and oxidation step with a scan rate of 50 mV s$^{-1}$:$E_{pc} = -0.60$ V vs. SCE in cathodic region ($2C_{16}V^{2+}$ to $2C_{16}V^{+\bullet}$) and $E_{pa} = -0.40$ V vs. SCE in anodic region ($2C_{16}V^{+\bullet}$ to $2C_{16}V^{2+}$). Peak top currents in both cathodic and anodic regions were largely influenced by temperatures (fluidity of the bilayer film). Figure 12 shows the temperature dependence of cathodic ($I_{pc}$) and anodic

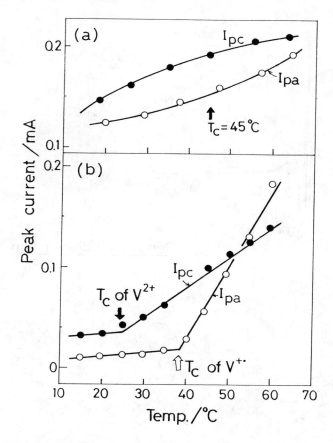

FIGURE 12.    Temperature dependences of cathodic ($I_{pc}$) and anodic peak currents ($I_{pa}$) of (a) the $2C_{12}SO_3^-$/polyV$^{2+}$ film and (b) the $2C_{16}V^{2+}$/PSS$^-$ film. Phase transition temperatures ($T_c$) of the bilayer film were separately obtained by DSC measurements.

peak currents ($I_{pa}$) of the $2C_{16}V^{2+}$/PSS$^-$ film, together with that of the $2C_{12}SO_3^-$/polyV$^{2+}$ film. The $I_{pc}$ and $I_{pa}$ of the $2C_{16}V^{2+}$/PSS$^-$ film increased drastically above 24 and 38°C, which are well consistent with phase transition temperatures of the $2C_{16}V^{2+}$ and $2C_{16}V^{+\cdot}$ bilayers, respectively. Thus, the reduction of $2C_{16}V^{2+}$ to $2C_{16}V^{+\cdot}$ films occurred only in the fluid state of $2C_{16}V^{2+}$ bilayers above $T_c = 24$°C, but not in the solid state below the $T_c$. Similarly, the oxidation of $2C_{16}V^{+\cdot}$ to $2C_{16}V^{2+}$ occurred only in the liquid crystalline state of $2C_{16}V^{+\cdot}$ bilayers above the $T_c = 38$°C. An explanation for this is that the mobility of the charge-compensating counter ions (ClO$_4^-$) is reduced in the solid state of bilayers compared with that in the fluid liquid crystalline bilayers.

On the contrary, the redox current peaks of the $2C_{12}SO_3^-$/polyV$^{2+}$ film, in which the viologen redox site was located in the hydrophilic polymer chain but not in the bilayer matrix, simply increased with temperatures. Thus, the mobility of counter anions may not be affected by the phase transition of bilayers. These findings indicate that electrochemical redox reactions of the viologen unit buried in lipid bilayer matrices are largely influenced by the fluidity of the bilayer film.

## B. PERMEABILITY CONTROL

Figure 13 shows typical time courses of the water-soluble nonionic fluorescent probe NQ$_2$ across the viologen-containing bilayer film cast on a Pt minigrid in N$_2$-purged 0.1 *M* NaClO$_4$ aqueous solution. The permeability was relatively fast (P = $3.0 \times 10^{-7}$ cm$^2$ s$^{-1}$)

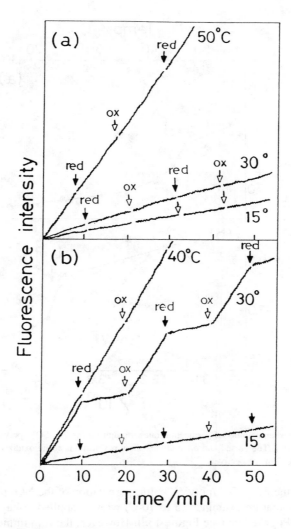

FIGURE 13. Permeation changes of the NQ$_2$ probe through (a) the 2C$_{12}$SO$_3^-$/polyV$^{2+}$ film and (b) the 2C$_{16}$V$^{2+}$/PSS$^-$ film cast on a Pt minigrid by electrochemical redox reactions in 0.1 $M$ NaClO$_4$ aqueous solutions. The film thickness is 100 $\mu$m. The potential of $-0.5$ and 0 V vs. SCE was applied to the Pt grid/film at red and ox, respectively.

in the oxidized form of the 2C$_{16}$V$^{2+}$/PSS$^-$ film. At the temperature of 30°C, when the potential of $-0.50$ V vs. SCE was applied to the Pt grid/film, the film turned from pale yellow to dark blue within 30 s, which shows the reduction of the dicationic 2C$_{16}$V$^{2+}$ to the radical cationic 2C$_{16}$V$^{+\cdot}$ and then the permeability decreased by a factor of 5 (P = 6.3 $\times$ 10$^{-8}$ cm$^2$ s$^{-1}$). When the potential on the Pt grid/film was switched off to 0 V even after along duration (10 min), the blue 2C$_{16}$V$^{+\cdot}$/PSS$^-$ film was oxidized again to the yellow 2C$_{16}$V$^{2+}$ film in 30 s, and the permeability reverted to the original fast rate (P = 3.3 $\times$ 10$^{-7}$ cm$^2$ s$^{-1}$).

At the temperature of 40°C, the permeation of the NQ$_2$ probe was hardly affected by the $-0.5$ to $-0.7$ V potential on the Pt grid/film in contrast to 30°C, although the redox reaction of the viologen film was confirmed to occur on the Pt grid from color changes. At 15°C, the redox reaction of the film and the permeation change were hardly observed in spite of an applied voltage of $-0.5$ to $-0.7$ V vs. SCE on the Pt grid/film. Thus, the reversible permeation change of the 2C$_{16}$V$^{2+}$/PSS$^-$ film responding to the redox reaction of viologen units was observed only in the range of around 25 to 35°C.

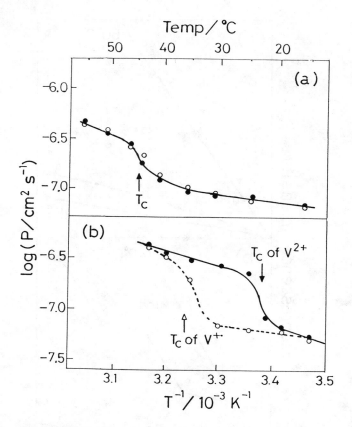

**FIGURE 14.** Arrhenius plots of permeation rates through (a) the $2C_{12}SO_3{}^-$/poly$V^{2+}$ film and (b) the $2C_{16}V^{2+}$/PSS$^-$ film. The closed and open circles show permeations at the oxidized ($V^{2+}$) and reduced form ($V^+$) of the film, respectively.

In the case of the $2C_{12}SO_3{}^-$/poly$V^{2+}$ film, permeation of the $NQ_1$ probe did not change at all over the temperature examined (5 to 70°C) with an applied voltage of $-0.5$ to $-0.7$ V vs. SCE on the Pt grid/film (see Figure 13a). However, the film immediately turned from yellow to dark blue with the voltage which means the smooth reduction of viologen units in the film all over the temperature. This implies that the redox reaction of viologenes located in the hydrophilic polymer chain, but not in the lipid bilayer matrix of the $2C_{12}SO_3{}^-$/poly $V^{2+}$ film, does not affect the permeability of the bilayer film.

## C. PERMEATION MECHANISM

For clarification of the effect of temperature on the permeability change of the $2C_{16}V^{2+}$/PSS$^-$ film, permeation rates of $NQ_2$ across the film both with and without the applied voltage of $-0.5$ V vs. SCE on the Pt grid at various temperatures (15 to 55°C) were obtained. Arrhenius plots are shown in Figure 14, together with the results of the $2C_{12}SO_3{}^-$/poly$V^{2+}$ film. The permeability of the oxidized $2C_{16}V^{2+}$/PSS$^-$ film without a potential on a Pt grid gave an abrupt inflection in the temperature range of 22 to 25°C. This inflection corresponded well to $T_c = 24$°C of the oxidized $2C_{16}V^{2+}$ bilayers obtained by DSC measurements. In the case of the radical cationic $2C_{16}V^{+\cdot}$/PSS$^-$ film reduced electrochemically by $-0.5$ V potential, the permeability jumped at the temperature range of 35 to 40°C which is consistent with the phase transition of the reduced $2C_{16}V^{+\cdot}$ bilayers ($T_c = 38$°C).

The permeability of the $2C_{12}SO_3{}^-$/poly$V^{2+}$ film increased in the fluid state above $T_c = 45$°C relative to the solid state of bilayers, although the inflection point ($T_c$) was not

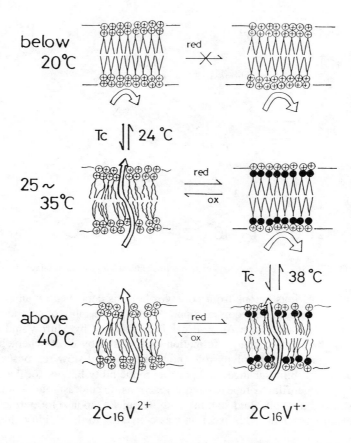

**FIGURE 15.** A schematic illustration of the permeation mechanism of the $2C_{16}V^{2+\cdot}/PSS^-$ film depending on both electrochemical redox reactions and temperatures.

changed by the electrochemical redox of viologen units in the film (Figure 14a). This means that the permeability of the $2C_{12}SO_3^-/polyV^{2+}$ film changed simply with the phase transition of lipid bilayer matrices, independent of the redox reaction in polymer chains.

The redox-sensitive permeation of the $2C_{16}V^{2+}/PSS^-$ film depending on temperature can be examined as follows according to the schematic illustration in Figure 15. At temperatures of 25 to 35°C, the oxidized $2C_{16}V^{2+}$ bilayers are in the fluid liquid crystalline state above $T_c = 24°C$. When the film was reduced to the $2C_{16}V^{+\cdot}/PSS^-$ form, the fluid state of the $2C_{16}V^{2+}$ bilayers changed to the rigid solid state of the $2C_{16}V^{+\cdot}$ bilayers because the phase transition temperature of the $2C_{16}V^{2+}$ bilayers ($T_c = 24°C$) increased to $T_c = 38°C$ of the $2C_{16}V^{+\cdot}$ bilayers, which have a high resistance to the permeation. In contrast, above 40°C, both $2C_{16}V^{2+}$ and $2C_{16}V^{+\cdot}$ bilayers are in the fluid state above their $T_c$ and showed a similar high permeability independent of their redox forms. At temperatures below 20°C, both $2C_{16}V^{2+}$ and $2C_{16}V^{+\cdot}$ bilayers are in the solid state below their $T_c$ and the redox reaction of viologen units hardly occurred in the solid state of the bilayers, as indicated in cyclic voltammograms of Figure 12b, and then the permeability does not change. Thus, the permeability changes responding to redox reactions of the $2C_{16}V^{2+}/PSS^-$ film at 25 to 35°C should be attributed to the fluidity change of the bilayer film because of the $T_c$ transition of bilayers by redox reactions.

The permeability of the $2C_{12}SO_3^-/polyV^{2+}$ film, in which viologen units exist in polymer chains but not in bilayer matrices, was not affected by the redox reaction (see Figures 13a and 14a). The change of membrane charges by redox reactions should be unimportant for the permeation of nonionic probes.

$$2C_{16}N^+Fc/PSS^- \qquad\qquad FcC_{11}C_{18}N^+/PSS^-$$

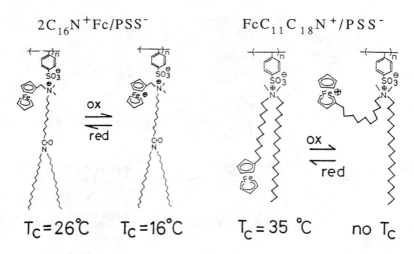

$$T_C = 26°C \qquad T_C = 16°C \qquad\qquad T_C = 35\ °C \qquad no\ T_C$$

FIGURE 16.   Ferrocene-containing electrochemical-sensitive lipid bilayer films.

The release of various anions from polypyrrole films deposited on an electrode can be changed by the electrochemical redox reaction.[60-63] The permeability of chloride anion through polypyrrole films deposited on a Au grid has been reported by Murray and co-workers to be increased by the positive charge formation on the film by electrochemcial redox reactions.[60] This ion gate film of polypyrrole, however, showed slow responses, taking 15 to 30 min for the permeability change, and poor reproducibility because of the morphological damages of the polymer film by repeated redox reactions. In contrast, our viologen-containing bilayer film showed the high fluid stability and the quick response for permeation changes, because the electrochemical redox reaction was converted to the fluidity change of bilayer films.

### D. FERROCENE-CONTAINING FILMS[43,44,50]

We have prepared another type of redox-sensitive film containing a ferrocene unit as a redox site at a hydrophobic alkyl chain ($FcC_{11}C_{18}N^+/PSS^-$) or at a hydrophilic head group ($2C_{16}N^+Fc/PSS^-$) (Figure 16) and studied the permeation control responding to electrochemical reactions.[50] Redox reactions of a ferrocene group in multibilayer films ($FcC_{11}C_{18}N^+/PSS^-$ and $2C_{16}N^+Fc/PSS^-$) occurred only in the fluid state of bilayer matrices above their phase transition temperatures ($T_c$ = 35 and 26°C, respectively) by cyclic voltammograms. Permeability of water-soluble nonionic fluorescent probes through the multibilayer film cast on a platinum minigrid sheet enhanced by a factor of 2 to 5 by an electrochemical oxidation of a ferrocenyl unit in the fluid multibilayer film above their $T_c$, compared with that of the reduced form. Permeability could be changed reversibly at least 20 cycles by repeated redox reactions in the case of the $2C_{16}N^+Fc/PSS^-$ film containing a Fc unit near the hydrophilic part. In contrast, the permeation change of the $FcC_{11}C_{18}N^+/PSS^-$ film having a redox site in the hydrophobic dialkyl chain became irreversible under continuous redox reactions.

When the ferrocene site was introduced in the side chain of synthetic polypeptide, the permeability through the cast film of copoly(γ-ferrocenylmethyl-L-glutamate/g-methyl-L-glutamate) could be controlled by the electrochemical redox reactions due to the peptide conformation changes from α-helix to random coil in the film.[43,44]

## VI. SUMMARY

In this review, we showed the multibilayer-immobilized cast films can control reversibly their permeability responding to various outside effects such as temperature, ambient pH, electric field, and electrochemical reactions. The fluid lipid layers act as a permeation valve. These signal-receptive permeability-controllable lipid membranes will become a new tool for the study of transport mechanisms of biological membranes.

# REFERENCES

1. **Szoka, F. and Papahadjopourous, D.,** *Liposomes: From Physical Structures to Therapeutic Applications,* Elsevier, Amsterdam, 1981.
2. **Mortonosi, A.,** *Membrane and Transport,* Plenum Press, New York, 1982.
3. **Fendler, J. H.,** *Membrane Mimetic Chemistry,* Wiley Interscience, New York, 1982.
4. **Kunitake, T. and Okahata, Y.,** A totally synthetic bilayer membrane, *J. Am. Chem. Soc.,* 99, 3860, 1977.
5. **Kunitake, T.,** Chemistry of synthetic bilayer membranes, *J. Makromol. Sci. Chem.,* A13, 587, 1979.
6. **Fendler, J. H.,** Surfactant vesicles as membrane mimetic agents. Characterization and utilization, *Acc. Chem. Res.,* 13, 7, 1980.
7. **Murakami, Y., Nakano, A., Yoshimatsu, A., and Fukuya, K.,** Functionized vesicular assembly. Enantioselective catalysis of ester hydrolysis, *J. Am. Chem. Soc.,* 103, 728, 1981.
8. **Deguchi, M. and Mino, J.,** Solution properties of long-chain dialkyldimethylammonium salts. 1. Formation of vesicles by dioctadecyldimethylammonium chloride, *J. Colloid. Interface Sci.,* 65, 155, 1978.
9. **Riberio, A. M. and Chaimovich, H.,** Preparation and characterization of large dioctadecyldimethylammonium chloride liposomes and comparison with small sonicated vesicles, *Biochim. Biophys. Acta,* 733, 172, 1983.
10. **Sudhoelter, E. J. R., Grip, W. J., and Engberts, J. B.,** Rhodopsin reconstitution in vesicles formed from simple, fully synthetic amphiphiles, *J. Am. Chem. Soc.,* 104, 1069, 1982.
11. **Fuhrhop, J-H., Fritsch, D., Tesche, B., and Schmiady, H.,** Water-soluble a,ω-bis(paraquat) amphiphiles from monolayer membrane vesicles, micellar, and crystals by stepwise anion exchange or photochemical reduction, *J. Am. Chem. Soc.,* 106, 1998, 1984.
12. **Moss, R. A. and Ihara, Y.,** Cleavage of phosphate esters by hydroxyl-functionized micellar and vesiclar reagents, *J. Org. Chem.,* 48, 588, 1983.
13. **Hains, T. H.,** Anionic lipid headgroups as a proton-conducting pathway along the surface of membranes: A hypothesis, *Proc. Natl. Acad. Sci. U.S.A.,* 80, 160, 1983.
14. **Fendler, J. H.,** Polymerized surfactant vesicles: Novel membrane mimetic systems, *Science (Washington, D.C.),* 223, 888, 1984.
15. **Kunitake, T., Nakashima, N., Takarabe, K., Nagai, M., Tsuge, A., and Yanagi, M.,** Vesicles of polymeric bilayer and monolayer membranes, *J. Am. Chem. Soc.,* 103, 5945, 1981.
16. **Elbert, R., Laschewsky, A., and Ringsdorf, H.,** Hydrophilic spacer groups in polymerizable lipids: Formation of biomembrane models from polymerized lipids, *J. Am. Chem. Soc.,* 107, 4134, 1985.
17. **Lopez, E., O'Brien, D. F., and Whitesides, T. H.,** Structural effects on the photopolymerization of bilayer membranes, *J. Am. Chem. Soc.,* 104, 305, 1982.
18. **Regen, S. L., Singh, A., Oehme, G., and Singh, M.,** Polymerized phosphatidylcholine vesicles. Synthesis and characterization, *J. Am. Chem. Soc.,* 104, 791, 1982.
19. **Nishide, H. and Tsuchida, E.,** Synthesis of novel styrene groups containing glycerophosphocholines and their polymerization, *Makromol. Chem. Rapid Commun.,* 5, 779, 1984.
20. **Regen, S. L., Shin, J-S., and Yamaguchi, K.,** Polymer-encased vesicles, *J. Am. Chem. Soc.,* 106, 2446, 1984.
21. **Iwamoto, K. and Sunamoto, J.,** Liposomal membranes. XII. Adsorption of polysaccharides on liposomal membranes as monitored by fluorescence depolarization, *J. Biochem. (Tokyo),* 91, 975, 1982.
22a. **Okahata, Y.,** Lipid bilayer-corked capsule membranes. Reversible, signal-receptive permeation control, *Acc. Chem. Res.,* 19, 57, 1986.
22b. **Okahata, Y.,** Signal-receptive capsule membranes. Permeability control of capsule membranes having bilayer-corkings or surface-grafted polymers as a permeation valve, in *Contemporary Topics in Polymer Science,* Vol. 2, Ottenbrite, R. M., Utracki, L. A., and Inoue, S., Eds., Hauser Pub., New York, 1987, chap. 6.5.
23. **Okahata, Y., Hachiya, S., Ariga, K., and Seki, T.,** The electrical breakdown and permeability control of a bilayer-corked capsule membrane in an external electric field, *J. Am. Chem. Soc.,* 108, 2863, 1986.
24. **Okahata, Y., Noguchi, H., and Seki, T.,** Functional capsule membranes. 26. Permeability control of polymer-grafted capsule membranes responding to ambient pH changes, *Macromolecules,* 20, 15, 1987.

25. **Okahata, Y. and Ariga, K.,** A capsule membrane grafted with viologen-containing polymers as a reactor of electron-transfer catalysis in heterophases, *J. Chem. Soc.,* Perkin Trans., 2, 1003, 1987.

26. **Okahata, Y., Nakamura, G., and Noguchi, H.,** Functional capsule membranes. 29. Concanavalin A-induced permeability control of capsule membranes corked with the synthetic glycolipid bilayers or grafted with synthetic glycopolymers, *J. Chem. Soc., Perkin Trans.,* 2, 1317, 1987.

27. **Okahata, Y. and Ijiro, K.,** Functional capsule membranes. 29. Thermolysin-immobilized capsule membranes as bioreactors in the synthesis of a dipeptide (precursor of aspartame in an organic solvent, *J. Chem. Soc., Perkin Trans.,* 2, 91, 1988.

28. **Okahata, Y., Ariga, K., and Seki, T.,** Polymerizable lipid-corked capsules. Polymerization at different positions of corking lipid bilayers on the capsule and effect of polymerization on permeation behavior, *J. Am. Chem. Soc.,* 110, 2495, 1988.

29. **Kajiyama, T., Kumano, A., Takayanagi, M., Okahata, Y., and Kunitake, T.,** Crystal-liquid crystal phase transformation and water permeability of artificial amphiphiles as biomembrane model, *Chem. Lett.,* 645, 1979.

30. **Kumano, A., Kajiyama, T., Takayanagi, M., Kunitake, T., and Okahata, Y.,** Phase transition behavior and permeation properties of cationic and anionic artificial lipids with two alkyl chains, *Ber. Bunsenges. Phys. Chem.,* 88, 1216, 1984.

31. **Kumano, A., Niwa, O., Kajiyama, T., Takayanagi, M., and Kunitake, T.,** Photoresponsive permeation characteristics of a ternary composite membrane of polymer/artificial lipid/azobenzene derivatives, *Polym. J.,* 16, 461, 1984.

32. **Shimomura, M. and Kunitake, T.,** Immobilization of synthetic bilayer membranes as multibilayered polymer films, *Polym. J.,* 16, 187, 1984.

33. **Higashi, N. and Kunitake, T.,** Immobilization of cast bilayer films by $^{60}$Co $\gamma$-irradiation and other means, *Polym. J.,* 16, 584, 1984.

34. **Hayashida, S., Sato, H., and Sugawara, S.,** Bilayer structures of artificial lipids in a water-soluble polymer binder, *Chem. Lett.,* 625, 1983.

35. **Okahata, Y., Lim, H-J., Nakamura, G., and Hachiya, S.,** A large nylon capsule coated with a synthetic bilayer membranes. Permeability control of NaCl by phase transition of the dialkylammonium bilayer coatings, *J. Am. Chem. Soc.,* 105, 4855, 1983.

36. **Sada, E., Katoh, S., and Terashima, M.,** Permeabilities of composite membranes containing phospholipids, *Biotech. Bioeng.,* 25, 317, 1983.

37. **Araki, K., Kon-no, R., and Seno, M.,** Gas permeability of phosphatidylcholine impregnated in a porous cellulose nitrate membrane, *J. Membr. Sci.,* 17, 89, 1984.

38. **Higashi, N., Kahiyama, T., Kunitake, T., Plass, W., Ringsdorf, H., and Takahara, A.,** Cast multibilayer film from polymerizable lipids, *Macromolecules,* 20, 29, 1987.

39. **Nakashima, N., Kunitake, M., Kunitake, T., Tone, S., and Kajiyama, T.,** Ordering cast films of polymerized bilayer membranes, *Macromolecules,* 18, 1515, 1985.

40. **Kunitake, T., Tsuge, A., and Nakashima, N.,** Immobilization of ammonium bilayer membranes by complexation with anionic polymers, *Chem. Lett.,* 1783, 1984.

41. **Okahata, Y. and En-na, G.,** Electric responses of bilayer-immobilized films as a model of chemoreceptive membranes, *J. Chem. Soc., Chem. Commun.,* 1365, 1987.

42. **Okahata, Y., Ebato, H., and Taguchi, K.,** Specific adsorption of bitter substances onto lipid bilayer-coated piezoelectric crystals, *J. Chem. Soc., Chem. Commun.,* 1363, 1987.

43. **Okahata, Y. and Takenouchi, K.,** Reversible permeability control of polypeptide membrane by electrochemical redox reactions in side chain ferrocene groups, *J. Chem. Soc., Chem. Commun.,* 558, 1986.

44. **Okahata, Y. and Takenouchi, K.,** Permeability controllable membranes. 9. Electrochemical redox-sensitive gate membranes of polypeptide films having ferrocene groups in side chain, *Macromolecules,* 22, 308, 1989.

45. **Okahata, Y. and Shimizu, A.,** Preparation of bilayer-intercalated cray films and permeation control responding to temperatures, electric fields, and ambient pH changes, *Langmuir,* 5, 954, 1989.

46. **Okahata, Y., Taguchi, K., and Seki, T.,** Electrically modulated permeability control of bilayer films, *J. Chem. Soc., Chem. Commun.,* 1122, 1985.

47. **Okahata, Y., Fujita, S., and Iizuka, N.,** Bilayer-immobilized films containing mesogenic azobenzene amphiphiles, *Angew. Chem. Int. Ed. Engl.,* 25, 751, 1986.

48. **Okahata, Y., En-na, G., Taguchi, K., and Seki, T.,** Electrochemical permeability control through a bilayer-immobilized film containing redox sites, *J. Am. Chem. Soc.,* 107, 5300, 1985.

49. **Okahata, Y. and En-na, G.,** Permeability controllable membranes. 7. Electrochemical responsive gate membranes of a multibilayer film containing a viologen group as redox sites, *J. Phys. Chem.,* 92, 4546, 1988.

50. **Okahata, Y., En-na, G., and Takenouchi, K.,** Permeability controllable membranes. 8. Electrical redox sensitive permeation through a multibilayer-immobilized film containing a ferrocenyl groups as a redox site, *J. Chem. Soc., Perkin Trans.,* 2, 835, 1989.

51. **Zimmermann, U., Scheurich, P., Pilwat, G., and Benz, R.,** Cells with manipulated functions: New perspective for cell biology, medicine, and technology, *Agnew. Chem. Int. Ed. Engl.,* 20, 325, 1981.

52. **Zimmermann, U., Shultz, J., and Pilwat, G.,** Transcellular ion flow in *Escherichila Coli B* and electrical sizing of bacterias, *Biophys. J.,* 13, 1005, 1973.

53. **Teissie, J. and Tsong, T. Y.,** Electric field induced transient pores in phospholipid bilayer vesicles, *Biochemistry,* 20, 1548, 1981.

54. **Bhowmik, B. B. and Nandy, P.,** Electrical conductivity of bilayer lipid membranes in electrolytic solution, *Chem. Phys. Lipids,* 34, 101, 1983.

55. **Chernomordik, L. V., Sukharev, S. I., Abidor, I. G., Chizmadzhev, and Yu, A.,** Breakdown of lipid bilayer membranes in an electric field, *Biochim. Biophys. Acta,* 736, 203, 1983.

56. **Weiss, A.,** Organic derivatives of mica-type layer-silicates, *Angew. Chem. Int. Ed. Engl.,* 2, 134, 1963.

57. **Lagaly, G.,** Replication and evolution in Inorganic systems, *Angew. Chem. Int. Ed. Engl.,* 20, 850, 1981.

58. **Lagaly, G. and Weiss, A.,** Experimental evidence for kink formation, *Angew. Chem. Int. Ed. Engl.,* 10, 558, 1971.

59. **Yamanaka, S., Matsunaga, M., and Hattori, M.,** Preparation and structure of n-alkyl ester derivatives of zirconium bis-(monohydrogen orthophosphate)dihydrorate with a layer structure, *J. Inorg. Nucl. Chem.,* 43, 1343, 1981.

60. **Burgmayer, P. and Murray, R. W.,** An ion gate membrane: Electrochemical control of ion permeability through a membrane with an embedded electrode, *J. Am. Chem. Soc.,* 104, 6139, 1982.

61. **Shinohara, H., Aizawa, M., and Shirakawa, H.,** Ion-sieving of electrosynthesized polypyrrole films, *J. Chem. Soc., Chem. Commun.,* 87, 1986.

62. **Shimizu, T., Ohtani, A., Iyoda, T., and Honda, K.,** Effective adsorption-desorption of cations on a polypyrrole-polymer anion composite electrode, *J. Chem. Soc., Chem. Commun.,* 1415, 1986.

63. **Miller, L. L., Zinger, B., and Zhou, Q-X.,** Electrically controlled release of $Fe(CN)_6^{4-}$ from polypyrrole, *J. Am. Chem. Soc.,* 109, 2267, 1987.

Chapter 7.5

# CENTRIFUGAL COUNTERCURRENT TYPE CHROMATOGRAPHY AS MULTISTAGE LIQUID MEMBRANE TRANSPORT SYSTEMS

Takeo Araki

## TABLE OF CONTENTS

# I. INTRODUCTION

In basic research laboratories, the selectivity efficiencies of carriers are usually tested by employing single-stage batch type liquid membrane transport (LMT) systems. As described in the preceding chapters, excellent carriers have been found for specific substrates. However, it will be quite difficult to find out the carriers having almost perfect selectivity for the majority of other substrates. Hence, in order to increase the separation efficiencies so that they are practical, multistage liquid membrane processes must be developed. The emulsion liquid membrane (ELM) process appears to be multistage in nature, and series arrangements of several unit modules can serve as a multistage system. Several multistage modules of supported liquid membrane (SLM) are already quite familiar. This chapter will concentrate on a third possible chromatographic multistage process, which is a new candidate for multistage operation of batch type liquid membrane transport systems.

# II. TYPES OF MULTISTAGE LIQUID MEMBRANE TRANSPORT SYSTEM

## A. WHY LIQUID MEMBRANE?

After lengthy fundamental investigations on material separation and concentration using the processes of mass transfer through liquid membranes, industrial application is now close to the goal. The most attractive feature of liquid membrane compared with solid polymer membrane is its extremely high mobility as a liquid, which leads to very rapid interfacial and in-phase diffusion. Pore-free permeation thus occurs quickly. Ultra-thin liquid films can be quite readily prepared and very small amounts of highly selective carrier compounds can be impregnated in the liquid membranes.

## B. EMULSION LIQUID MEMBRANE OR LIQUID SURFACTANT MEMBRANE

In the ELM processes there are cases where a substrate species permeates successively into many microemulsion globules while it is being transferred from the supply to the receiving vessel. This process thus seems to be multistage in nature. In 1966 Li of the Esso group patented a desalination process using ELM,[1] in 1971 Cussler achieved active transport of Na ion,[2] and in 1986 Marr and his collaborators constructed the first commercial scale plant for recovery and concentration of Zn ion from viscose-waste solution. Comprehensive reviews on the basic concept of the latter[3] and basic data for the plant have been reported.[4-5] One reason for the good efficiency of these processes may be the intrinsic multistage nature of a flux through microglobules of the emulsion. The dynamic molecular motions in a surfactant layer of an emulsion globule cause instability of the emulsion. Extensive efforts to overcome drawbacks in the ELM processes are now being made by several research groups.

## C. SUPPORTED LIQUID MEMBRANE

Another promising multistage liquid membrane process is the SLM system, in which liquid membranes are impregnated in the micropores and/or thin portions of a polymer hollow fiber and/or polymer sheet. Use of porous cellulose acetate as the supporting material for an immobilized film of water was tested in the 1960s to accelerate the transport of $O_2$,[6] for desalination[7] and then for $CO_2$-$O_2$ separation.[8] The SLM process permits the use of physically stable ultra-thin liquid films and the ready construction of multistage modules, as in the solid membrane process and is also still under investigation.

## D. CHROMATOGRAPHY RELATED TO LIQUID MEMBRANE TRANSPORT

Although mathematical treatment for the facilitated mass transport in liquid membrane systems using planar sheet models can basically and quantitatively predict their operating results as shown by Ward[9] and Cussler,[2] more work with sphere and cylinder models is required for the ELM and SLM processes, respectively.[10b] The two types of processes also involve severe inherent drawbacks for practical operation, mainly due to problems of operational stability which prevent their smooth development for industrial use.[10-12]

The ELM processes are often operated using mixer-settler type continuous extraction devices.[13-15] Other apparatus for the continuous process has also been designed to develop certain chromatographic multistage effects. The SLM process can be operated in modules thereby resembling liquid chromatography. However, this has rarely been referred to as liquid membrane chromatography, because the net separation efficiency for multi-component fractionation is much less than familiar chromatography.

Chromatography is a technique in which the effect of a unit process is amplified by thousands of repetitions, e.g., recent high performance chromatographies permit complete separation even if the separation factor of two given samples is >1.1 Liquid membranes coated on the particles of solid porous support have been used in gas-liquid partition chromatography, but this method is not covered in this chapter.

In the 1950s, Craig's classical continuous multistage extraction process called the "countercurrent distribution method",[16] an original form of liquid-liquid two phase extraction chromatography, was in common use in biochemical laboratories, and a great quantity of separation/isolation data were compiled.[17-22] This classical process, however, was quite complicated and extremely time-consuming.

Recently, Craig's method has been greatly modernized for rapid and easy operation by applying centrifugal force.[23-25] Two types of chromatographic machines are now commercially available for laboratory use, and excellent fractionation/isolation/enrichment results have been obtained in bioorganic and other laboratories.[26-43] Industrial scale chromatographic apparatus is also available. The fundamental theory of modern centrifugal counter-current type chromatographies is almost identical to that established by the classical method.[17] Only minor modifications are necessary on the effects of centrifugal force when maximum efficiency is desired.[25-27,45-46]

Although centrifugal countercurrent type chromatography in the standard form (only partition effect between two-solvents) is not directly correlated with the facilitated LMT systems, Araki et al.[47] showed that when a carrier or a separator is impregnated in the stationary phase, the separation behaviors become closely related to the facilitated LMT results. Hence, the new chromatography, especially "separator- (or carrier-)aided" centrifugal countercurrent type chromatography, seems to be a third promising possibility of multistage liquid membrane transport process for separation/isolation/purification. The following sections will focus on this subject.

## III. TYPES OF CENTRIFUGAL COUNTERCURRENT EXTRACTION METHODS

### A. CENTRIFUGAL PARTITION CHROMATOGRAPHY

Centrifugal partition chromatography (CPC) was introduced by Murayama et al. of the Sanki Engineering group in 1982.[25] The machinery characteristics of CPC are the simple and compact construction of microcell-cartridges and the use of rotary seal joints. In the handy laboratory scale equipment, 50 or 400 microcells are arranged in series in a Teflon cartridge block (ca. $50 \times 40 \times 200$ mm), and a maximum of 12 cartridges (containing maximum 4,800 microcells) can be set in series on a rotary disk. The basic features of the unit microcell are depicted in Figure 1a. Upon rotation of the disk, one of the two immiscible liquid phases acts as the stationary phase as a result of the centrifugal force applied. The researchers can select an optimum from four types of elution modes, i.e., organic eluent/aqueous stationary phase (ascending or descending) or aqueous eluent/organic stationary phase (ascending or descending). A schematic sketch of the cartridge construction of the rotary disk is depicted in Figure 1b.

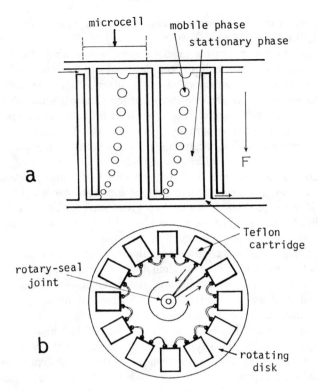

FIGURE 1.   The basic processes of centrifugal countercurrent type chromatographies. (a) A unit microcell in CPC.[25] (b) Cartridges containing microcells are arranged on a rotary disk in CPC.[25]

The machinery construction of CPC is closely related to that of Craig's continuous extraction method, i.e., a series arrangement of extraction compartments. Hence, Craig's distribution theory can be almost directly applied, permitting estimation of the ideal number of theoretical plates quite readily.

## B. CENTRIFUGAL COUNTERCURRENT CHROMATOGRAPHY

The present style of high speed centrifugal countercurrent chromatographic apparatus (CCC) was developed by Ito of NHLBI somewhat earlier than CPC.[23c] This was an epoch change from the slow countercurrent chromatography, including droplet countercurrent chromatography (DCCC), to today's high-speed techniques. Ito's apparatus is called a "coil planet centrifuge" because a coiled tube(s) is rotated outside a rotating diaphragm (Figure 2b) to attain the unilateral hydrodynamic equilibrium in a tube (Figure 2a). This apparatus is operated without either rotary seal joint or microcell compartments. Again, researchers can choose the optimum operational mode as in the case of CPC. However, estimation of the ideal number of theoretical plates is rather difficult in this noncompartment system. Length and combination of the tube(s) seem to be the user's empirical choice. Several mechanical modifications have been advanced by Ito's group.

## IV. CORRELATION BETWEEN CENTRIFUGAL PARTITION CHROMATOGRAPHY AND LIQUID MEMBRANE TRANSPORT SYSTEM

### A. BASIC CONCEPT

A single unit behavior in "separator- (or carrier-)aided" CPC (CCC has basically similar unit behavior) is illustrated in Figure 3a in comparison with that of a single stage facilitated

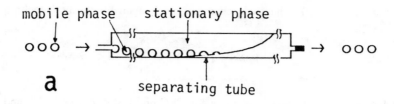

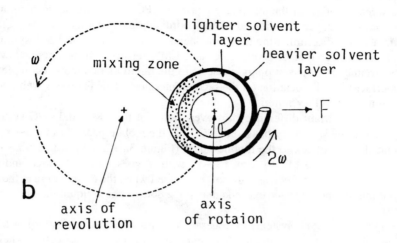

FIGURE 2.   Schematic sketches of the mechanical constructions of centrifugal countercurrent type chromatographies. (a) A separating tube in CCC.[26] (b) Coil planet centrifuge apparatus (CCC) in which the coiled tube(s) is rotated outside a rotating diaphragm.[26]

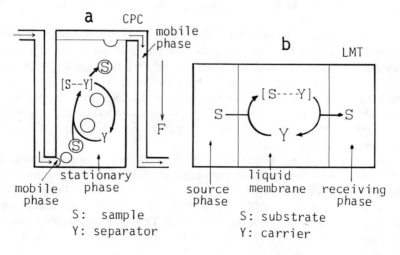

FIGURE 3.   Unit behaviors in the "separator- (or carrier-)aided" CPC (a), and the single stage facilitated LMT systems (b).

LMT system (Figure 3b). The basic chemical mechanism for the both techniques can be described as[48-49]

$$S + Y \rightarrow [S\text{----}Y] \rightarrow Y + S \qquad (1)$$

where S denotes a sample and a substrate for CPC and LMT, respectively, and Y denotes a separator and a carrier for CPC and LMT systems, respectively. [S----Y] represents interactions between S and Y in the stationary phase in CPC and in the liquid membrane. The [S----Y] includes hydrogen bonding, dipolar, coordinating, chelating, and any other interactions effective for discriminating the components Ss in the CPC and LMT systems. From Equation 1, it is clear that effective CPC and LMT results will be obtained when Y is effectively circulated in the stationary phase or in the liquid membrane via complexation (at the first interface) and dissociation (at the second interface). This feature is well documented in the literature of LMT systems.

There is an important difference, however, between the LMT and CPC systems, i.e., S transferred into the stationary phase in CPC from the mobile phase forms [S----Y] and then the dissociated S goes back to the mobile phase (Figure 3a). In contrast, S transferred into the liquid membrane in LMT to form [S----Y] with Y goes into the source and receiving phases after dissociation. Effective transport is attained when the dissociation at the receiving phase/liquid membrane interface exceeds that at the source phase/liquid membrane interface (Figure 3b).

In the facilitated LMT system, in general, Y is necessary for transporting S from the source phase to the liquid membrane. The separator-aided CPC is effective when a set of samples $S_1$ and $S_2$ has the same retention volume ($V_R$) in the normal partition type CPC without Y. In the separator-aided CPC, $V_R = 0$ is not a necessary condition, but Y in the stationary phase is solely responsible for production of the difference in $V_R$ values between $S_1$ and $S_2$. In a typical case where the [S----Y] is a very weak interaction, S will have a very small $V_R$ value in CPC. In the corresponding LMT system, however, the transport will require rather a long time.

Thus, in spite of their chemical similarity (Equation 1), the observed behaviors in LMT and CPC will not always be parallel due to their physical differences. To conventionally compare separator-aided CPC with the corresponding facilitated LMT system, it is assumed that $S_1$ and $S_2$ cannot permeate into either the stationary phase or the liquid membrane in the absence of Y, and that the [S----Y] interaction is the sole factor causing the difference in the transfer behaviors of $S_1$ and $S_2$. Under these assumptions one can obtain a rough guideline (Table 1)[48,50] of the results of CPC and the corresponding LMT by considering the ease of phase transfers involved.

For simplification, Table 1 shows only the case where Y is impregnated in the organic phase of a w/o/w two-liquid phase system in which S is transferred from w to o to w. In principle, the same pattern is applied for an o/w/o type system. [If Table 1 is treated more exactly, the term "ready" or "difficult" should be expressed by equilibrium constants (and also diffusion parameters, if necessary).] Although Table 1 is only a rough guide, it is very helpful in predicting the behaviors in CPC from results obtained by the corresponding LMT, and vice versa.

## B. EXAMPLES

The crown-ether facilitated selective LMT of alkali metal picrates[51,52] is a key finding in modern usage of liquid membranes as host-guest molecular recognition processes. Originally the transport selectivity of alkali metal ions showed almost direct correlation with extraction results obtained in fundamental organic host-guest chemistry.

# TABLE 1
## Relation Between CPC (Aqueous Mobile and Organic Stationary Phases) and LMT (Aqueous Source and Receiving Phases with Organic Liquid Membrane) in View of Phase Transfer Processes.[48,50]

| | Phase transfer processes | | Expected results | |
| --- | --- | --- | --- | --- |
| | Aqueous to organic phase | Organic to aqueous phase | Retention time in CPC | Transport rate in LMT |
| A) | No | — | Very short | 0 |
| B) | Ready | Ready | Short | High |
| C) | Ready | Difficult | Long[a] | Low |
| D) | Difficult | Ready | Short[b] | Low |
| E) | Difficult | Difficult | Very broad | Very low |
| F) | Boundary[c] | — | —[d] | 0 |

[a] Tailing peak.

[b] Leading peak.

[c] Permeation of S occurs at the boundary interface layer only, i.e., no permeation into the inside of organic phases.

[d] The retention time varies depending on the relative partition coefficients between the two immiscible phases.

Recently, the system containing dibenzo-18-crown-6(Y)/alkali picrates(S) in CHCl$_3$ was applied to CPC containing $50 \times 3$ microcells (ideal number of theoretical plates = 150).[47] Although 10%-butan-1-ol was added to the CHCl$_3$ phase in CPC, the retention values obtained by CPC were shown to be well correlated with logK (Figure 4), where K is the equilibrium constant obtained by corresponding simple extraction experiments.[53] In addition, the separation factor ($V_R(K^+)/V_R(Na^+)$) observed in CPC was 1.99 (log K(K$^+$/Na$^+$) = 2.0), resulting in effective separation/purification of these ions from the equimolar mixture (Figure 5). Other binary mixtures having $V_R(M_1)/V_R(M_2)$ = 1.25 to 1.55 (Table 2) will be effectively purified if a larger number of microcells is used in the CPC. No sets of alkali picrates were able to separate in the CPC without the crown-ether. Thus, CPC of the crown-ether/alkali picrate system was found to be indirectly correlated with the corresponding facilitated LMT system.

Direct comparison of separator-aided CPC with the corresponding LMT system was first made for separation/isolation of picric acid (PA) and p-nitrophenol (NP).[49] In this case no separation in CPC and no transport in LMT were observed using CHCl$_3$ organic phases alone. However, addition of 5%-butan-1-ol (Y) to the organic phase gave rise to complete separation of the two samples in CPC ($t_R$(NP)/$t_R$(PA) = 5.4) ($t_R$: retention time), allowing ready isolation of NMR-spectroscopically pure components. Use of the same CHCl$_3$/5%-butan-1-ol liquid membrane resulted in selective transport of the pair of substrates ($k_{PA}/k_{NP}$ = 8.4) (k: the apparent rate constant of transport). In both techniques NP stayed for a longer period in the organic phases than PA.

Separation and purification of lanthanide ions are important in separation chemistry. Many approaches[54-62] using di(2-ethylhexyl)phosphoric acid (D2EHPA), tri-butyl phosphate (TBP), 2-ethylhexyl phosphonic acid mono-2-ethylhexyl ester (PC-88A), and related phosphorus compounds as extractant have been tested by techniques such as liq.-liq. extraction, column chromatography,[63] 20-stage mixer-settler method,[64] and extraction by ELM.[65-66] Industrial purification of lanthanides is now being carried out on the basis of the fundamental extraction data obtained.

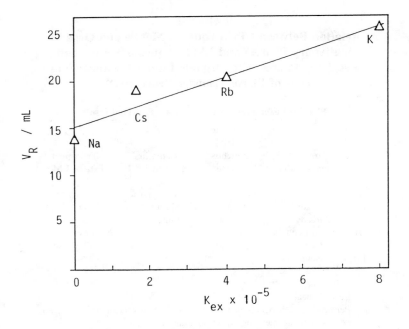

FIGURE 4. Plot of the $V_R$ values obtained by CPC vs. logK obtained[53] by the corresponding simple extraction experiments. Organic phase: dibenzo-18-crown-6 (Y) in $CHCl_3$/n-BuOH (9 + 1,v/v), and aqueous phase: $M^+$-picrates in $H_2O$.[47]

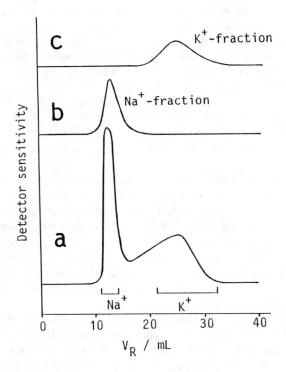

FIGURE 5. Chromatograms of CPC for separation of $Na^+$ and $K^+$ ions using dibenzo-18-crown-6 as Y in the 50 × 3 (= 150) micro cells.[47] (a) Equimolar binary mixture. (b) Chromatogram of the $Na^+$-fraction obtained by the first CPC(a). (c) Chromatogram of the $K^+$-fraction obtained by the first CPC(a). $t_R = t_R(apparent) - t_{dead}$.

## TABLE 2
### Comparison of CPC Separations with the Corresponding Extraction Equilibrium Constants for the System of Dibenzo-18-Crown-6(Y) in CHCl₃/Alkali Metal Picrates(S).[47]

| M⁺-picrates | | $V_R(M_1^+)/V_R(M_2^+)$ | | |
|---|---|---|---|---|
| $M_1^+$ | $M_2^+$ | Binary sample | Single samples | $\log(K_1/K_2)$[53] |
| Li⁺ | Na⁺ | 1.00 | 1.01 | — |
| K⁺ | Na⁺ | 1.99 | 1.86 | 2.0 |
| Rb⁺ | Na⁺ | 1.52 | 1.49 | 1.7 |
| Cs⁺ | Na⁺ | 1.50 | 1.38 | 1.3 |
| Cs⁺ | Li⁺ | 1.46 | 1.36 | — |
| K⁺ | Rb⁺ | 1.26 | 1.25 | 0.3 |

CPC of lighter rare earth metal ions (RECl₃ in aq. HCl) was performed using D2EHPA as separator in heptane.[67] Although the number of theoretical plates observed was low, good separations were obtained for the adjacent series of lighter RE ions from La to Sm by adjusting the HCl-concentration in the mobile phase (Figure 6). Parallelism between separation factors obtained by CPC with those by extraction was reasonable. Toluene which is more polar than heptane was an effective stationary solvent for the separation of Sm/Eu ions. Further study indicated that CHCl₃ (more polar than toluene) was effective for separations of the heavier RE ions such as an Er/Yb pair (Figure 6).[68,69]

The reason for peak broadenings is that the key reaction of RE ions with D2EHPA is a rather strong acid/base interaction (i.e., organic-to-aqueous transfer is difficult) as shown in Equation 2.[63] PC-88A was found to be an effective separator especially for heavier RE ions, when eluted with the aqueous mobile phase containing 20-vol% of ethylene glycol, which is buffered at a desired pH value with 0.1 $M$-(H,Na)Cl₂CHCOO (Figure 7).[70-71] Considerable peak broadening was also observed because the key reaction is the same type as Equation 2[70]:

$$3(HG)_2 + RE^{3+} \rightleftharpoons RE(GH_2)_3 + 3H^+ \qquad (2)$$

where G and RE denote phosphate anion and rare earth ions, respectively.

TBP was not an effective separator for RE(NO₃)₃ because its key reaction (Equation 3) is different from D2EHPA and is not adjustable by H⁺-concentration[68,69]:

$$RE^{3+} + 3NO_3^- + 3TBP \rightleftharpoons RE(NO_3)_3TBP \qquad (3)$$

For RE ions the direct comparison of CPC with the corresponding single-stage LMT system was studied using a D2EHPA/heptane liquid membrane and aq.-HCl phase.[50,69] The optimum HCl concentration for the transport becomes higher with an increase of the atomic number of REs (Figure 8), in parallel to the CPC. This parallelism was kept for the heavier RE members (e.g., see Reference 72a,b). Hence, this LMT system is found to be correlated with the corresponding CPC, and selective-LMT can be carried out by adjusting the HCl-concentrations. Namely, in applications to purification/isolation/enrichment the countercurrent type chromatographies will be of significant value as one type of multistage LMT system.

## V. OUTLOOK

Recent development of high speed CPC and CCC is a modernization of the continuous multistage liq.-liq. extraction method. When a suitable separator or carrier is impregnated

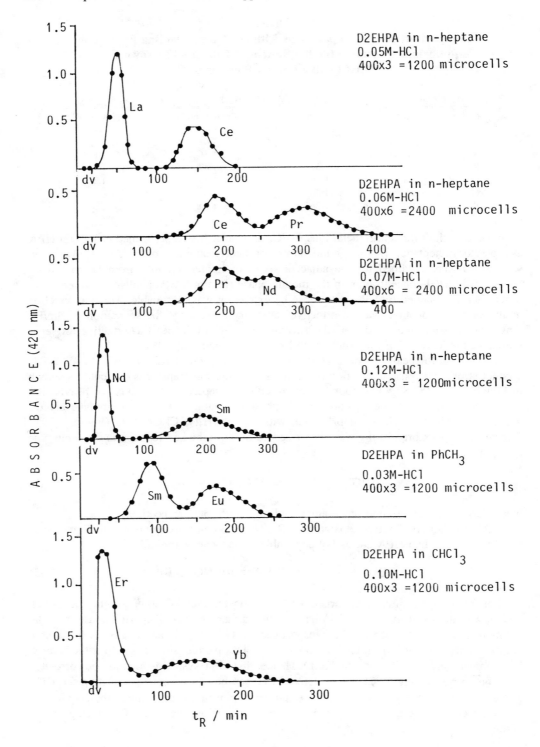

FIGURE 6.    CPC separations of RE ions by using D2EHPA as Y and eluted by aqueous HCl solution.[67-69] (dv: dead volume). The values $t_R$ include dead time.

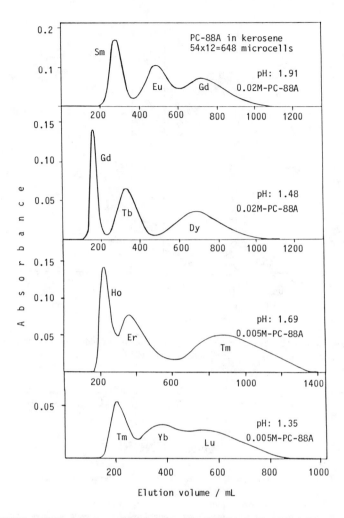

FIGURE 7.    CPC separations of RE ions by using PC-88A as Y and eluted by aqueous solution containing 20-vol. % of ethylene glycol buffered at a desired pH with 0.1 $M$-$(H,Na)Cl_2CHCOO$.[70] The values $t_R$ include dead time.

in the stationary liquid phase, the chromatographic behaviors have, in some cases, been shown to be correlated with the corresponding facilitated LMT systems. CPC and CCC can be relatively easily upgraded to an industrial scale. These chromatographic techniques will be further used in future complementally to the continuous ELM or SLM modules[3,72] which are advancing at present. Some of the severe problems now encountered in the ELM or SLM processes may be solved by the excellent multistage performances of CPC and CCC. In turn, since a large quantity of data, especially for biologically important molecules, has been accumulated during the long history of countercurrent distribution methods, that of a chromatographic nature will contribute to advancing of the fields of liquid membrane science and technology.

## ACKNOWLEDGMENTS

Our researches described were financially supported by the Ministry of Education, Science, and Culture with Grant-in-Aid for General Scientific Research (No. 61470026), and Mitsui Inter-business Research Institute. My thanks are also given to my collaborators, and to Sanki Engineering Co. Ltd.

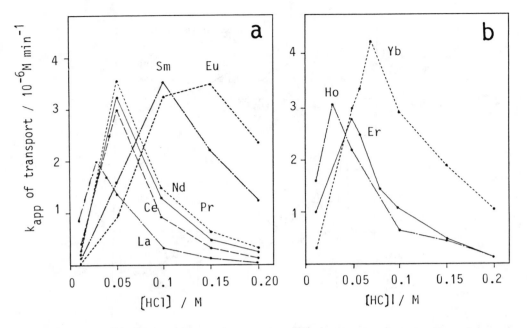

**FIGURE 8.** Single-stage LMT results for RE ions by using the corresponding liq.-liq. system to the CPC shown in Figure 6. Liquid membrane: (a) n-heptane containing D2EHPA as Y,[50,69] and (b) CHCl₃ containing D2EHPA as Y.[71]

# REFERENCES

1. **Li, N. N., U.S. Patent, 3,454,489, 1966; Li, N. N. and Somerset, N. J.,** Separating hydrocarbons with liquid membranes, U.S. Patent, 3,410,794, 1968.
2. **Cussler, E. L.,** Membranes which pump, *AIChE J.,* 17, 1300, 1971.
3. **Marr, R. and Kopp, A.,** Liquid membrane technology — a survey of phenomena, mechanisms, and models, *Chem. Ing. Tech.,* 52, 399, 1980; *Int. Chem. Eng.,* 22, 44, 1982.
4. **Draxler, J., Furst, W., and Marr, R.,** *Proc. ISEC'86,* 1, 1, 1986.
5. **Furst, W., Draxler, J., and Marr, R.,** *Proc. World Congr. III Chem. Eng.,* 3, 331, 1986.
6. **Scholander, P. F.,** Oxygen transport through hemoglobin solutions, *Science,* 131, 585, 1960.
7. **Loeb, S. and Sourirajan, S.,** Sea water demineralization by means of an osmotic membrane, *Advan. Chem. Ser.,* 38, 117, 1963.
8. **Ward III, W. J. and Robb, W. L.,** Carbon dioxide-oxygen separation: facilitated transport of carbon dioxide across a liquid film, *Science,* 156, 1481, 1967.
9. **Ward III, W. J.,** Analytical and experimental studies of facilitated transport, *AIChE J.,* 16, 405, 1970.
10a. **Noble, R. D. and Way, J. D.,** Liquid membrane technology. An overview, *ACS Symp. Ser., (Liquid Membrane),* 347, 1, 1987.
10b. **Way, J. D., Noble, R. D., Flynn, T. M., and Sloan, E. D.,** Liquid Membrane Transport: A survey, *J. Membr. Sci.,* 12, 239, 1982.
11a. **Izatt, R. M., Clark, G. A., Bradshaw, J. S., Lamb, J. D., and Christensen, J. J.,** Macrocycle-facilitated transport of ions in liquid membrane systems, *Sep. Pur. Meth.,* 15, 21, 1986.
11b. **Izatt, R. M., LindH, G. C., Bruening, R. L., Bradshaw, J. S., Lamb, J. D., and Christensen, J. J.,** Design of cation selectivity into liquid membrane systems using macrocyclic carriers, *Pure Appl. Chem.,* 58, 1453, 1986.
12. **Teramoto, M., Miyake, Y., and Matsuyama, H.,** Separation and concentration by liquid membranes, *Hyomen,* 25, 715, 1987.
13. **Jamrack, W. D., Logsdail, D. H., and Short, G. D. C.,** *Progress in Nuclear Energy,* Ser. 3, Process chemistry, Vol. 2, Bruce, F. R., Flentcher, J. M., and Hyman, H. H., Eds., Pergamon Press, New York, 1958, 320.

14. **Treybal, R. E.,** *Liquid extraction,* McGraw-Hill, New York, 1963.
15. **Rahn, R. W. and Smutz, M.,** Development of a laboratory scale continuous multistage extractor, *I&EC Proc. Des. Dev.,* 8, 289, 1969.
16. **Craig, L. C.,** Aconite alkaloids. XIII. The isolation of pimapthrene from the dehydrogenation products of stophisine, *J. Biol. Chem.,* 155, 519, 1944.
17. **Craig, L. C. and Craig, D.,** *Technique of Organic Chemistry,* Vol. 3, Interscience, New York, 1950.
18. **Hecker, E.,** *Verteilungsverfahren im Laboratorium,* Verlag Chemie, Berlin, 1955.
19. **Weisiger, J. R.,** *Organic Analysis,* Vol. 2, Interscience, New York, 1953.
20. **Rauen, H. M. and Stamm, W.,** *Gegenstromverteilung,* Springer Verlag, Berlin, 1953.
21. **Metzsch, F. A. V.,** The choice of solvents for the distribution between two liquid phases, *Angew. Chem.,* 65, 586, 1953.
22. **Tsugita, A.,** Koryu-bunpai ho, in *Jikken Kagaku Koza,* 2, Kisogijutsu II, Chem. Soc. Jpn., Maruzen, Tokyo, 1956, 303.
23a. **Ito, Y., Aoki, I., Kimura, E., Nunogaki, K., and Nunogaki, Y.,** New micro liquid-liquid partition techniques with the coil planet centrifuge, *Anal. Chem.,* 41, 1579, 1969.
23b. **Ito, Y. and Bowman, R. L.,** Countercurrent chromatography with flowthrough coil planet centrifuge, *Science,* 173, 429, 1971.
23c. **Ito, Y.,** Efficient preparative countercurrent chromatography with a coil planet centrifuge, *J. Chromatogr.,* 214, 122, 1981.
24a. **Ito, Y.,** Countercurrent chromatography, *J. Biochem. Bio. Phys. Method,* 5, 105, 1981.
24b. **Ito, Y., Sandlin, J., and Bowers, W. C.,** High-speed preparative countercurrent chromatography (CCC) with a coil planet centrifuge, *J. Chromatogr.,* 244, 247, 1982.
25. **Murayama, W., Kobayashi, T., Kosuge, Y., Yano, H., Nunobaki, Y., and Nunogaki, K.,** A new centrifugal counter-current chromatograph and its application, *J. Chromatogr.,* 239, 643, 1982.
26. **Ito, Y.,** High-speed countercurrent chromatography, in *CRC Critical Reviews in Analytical Chemistry,* Vol. 17, Issue 1, 65, 1986, and references cited therein.
27. **Mandava, N. B. and Ito, Y.,** *Countercurrent Chromatography: Theory and Practice,* Marcel Dekker, New York, 1987, and references cited therein.
28. **Sutherland, I. A., Heywood-Waggington, D., and Ito, Y.,,** Countercurrent chromatography. Applications to the separation of biopolymers, organelles and cells using either aqueous-organic or aqueous-aqueous phase systems, *J. Chromatogr.,* 384, 197, 1987.
29. **Bruening, R. C., Oltz, E. M., Furukawa, J., and Nakanishi, K.,** Isolation and structure of tunichrome B-1, a reducing blood pigment from the tunicate *Ascidia nigerra L., J. Am. Chem. Soc.,* 107, 5298, 1985.
30. **Okuda, T., Yoshida, T., Hatano, T., Yazaki, K., Kira, R., and Ikeda, Y.,** Chromatography of tannins. II. Preparative fractionation of hydrolyzable tannins by centrifugal partition chromatography, *J. Chromatogr.,* 362, 375, 1986.
31. **Okuda, T., Yoshida, T., and Hatano, T.,** Application of centrifugal partition chromatography to tannins and related polyphenols, *J. Liq. Chromatogr.,* 11, 2447, 1988.
32. **Jirousek, M. R. and Salomon, R. G.,** Purification of a water-sensitive natural product with an aprotic CPC solvent system, *J. Liq. Chromatogr.,* 11, 2507, 1988.
33. **Onji, Y., Aoki, Y., Yamazoe, Y., Dohi, Y., and Moriyama, T.,** Isolation of nivalenol and fusarenone-X from pressed barley culture, *J. Liq. Chromatogr.,* 11, 2537, 1988.
34. **Foucault, A. and Nakanishi, K.,** Study of AcOH/HCOOH/H$_2$O/CHCl$_3$ solvent systems. Application to the separation of two large and hydrophobic fragments of bacteriorhodopsin membrane protein by centrifugal partition chromatography, *J. Liq. Chromatogr.,* 11, 2455, 1988.
35. **Kusumoto, S., Kusunose, N., and Shiba, T.,** Chemical synthesis and purification of bacterial cell wall lipopolysaccharide, Proc. 2nd. CPC Colloq. (Progress and applications of CPC), June 4, Kyoto, 1988, 61.
36. **Lee, Y.-W., Cook, C. E., and Ito, Y.,** Dual countercurrent chromatography, *J. Liq. Chromatogr.,* 11, 37, 1988.
37. **Lee, Y.-W., Ito, Y., Fang, Q.-C., and Cook, C. E.,** Analytical high speed countercurrent chromatography, *J. Liq. Chromatogr.,* 11, 75, 1988.
38. **Knight, M., Pineda, J. D., and Burke, Jr., T. R.,** Solvent systems for the coutercurrent chromatography of hydrophobic neuropeptides analogs and hydrophilic protein fragments, *J. Liq. Chromatogr.,* 11, 119, 1988.
39. **Chen, R. H. and Hochlowski, J. E.,** Separation and purification of macrolides using the Ito multi-layer horizontal coil planet centrifuge, *J. Liq. Chromatogr.,* 11, 191, 1988.
40. **Mercado, T. I, Ito, Y., Strickler, M. P., and Ferrans, V. J.,** Countercurrent chromatography of an anti-trypanosomal factor from *Pseudomonas fluorescens, J. Liq. Chromatogr.,* 11, 203, 1988.
41. **Diallo, B. and Vanhaelen, M.,** Large scale purification of apocarotenoids from *Cochlospermum tinctorium* by countercurrent chromatography, *J. Liq. Chromatogr.,* 11, 227, 1988.

42. **Zhang, T.-Y., Hua, X., Xiao, R., and Kong, S.,** Separation of flavonoids in crude extract from sea buckthorn by countercurrent chromatography with two types of coil planet centrifuge, *J. Liq. Chromatogr.,* 11, 233, 1988.

43. **Freeman, H. S., Hao, Z., McIntosh, S. A., and Mills, K. P.,** Purification of some water soluble azo dyes by high-speed countercurrent chromatography, *J. Liq. Chromatogr.,* 11, 251, 1988.

44. **Murayama, W., Kosuge, Y., Nakaya, N., Nunogaki, Y., Nunogaki, K., Cazes, J., and Nunogaki, H.,** Preparative separation of unsaturated fatty acids esters by centrifugal partition chromatography (CPC), *J. Liq. Chromatogr.,* 11, 283, 1988.

45. **Armstrong, D. W.,** Theory and use of centrifugal partition chromatography, *J. Liq. Chromatogr.,* 11, 2433, 1988.

46. **Kosuge, Y., Murayama, W., and Nunogaki, Y.,** Systematization of centrifugal liquid-liquid partition chromatography, Proc. 1st. CPC Colloq. (Progress and applications of CPC), Kyoto, Jan. 13, 1986, 30.

47. **Araki, T., Kubo, Y., Toda, T., Takata, M., Yamashita, T., Murayama, W., and Nunogaki, Y.,** Carrier-aided centrifugal partition chromatography for preparative-scale separations, *Analyst,* 110, 913, 1985.

48. **Araki, T.,** Centrifugal countercurrent extraction method, *Kagaku,* 41, 512, 986.

49. **Araki, T., Kubo, Y., Takata, M., Gohbara, S., and Yamamoto, T.,** Selective liquid-membrane transport of nitrophenols by a simple and costless carrier. An application of the results of centrifugal partition chromatography, *Chem. Lett.,* 1987, 1011, 1987.

50. **Araki, T., Okazawa, T., Kobo, Y., Ando, H., and Asai, H.,** Liquid-membrane transport of rare earth metal ions facilitated by di(2-ethylhexyl)phosphoric acid: Comparison with the results of corresponding centrifugal partition chromatography, *J. Liq. Chromatogr.,* 11, 2487, 1988.

51. **Reusch, C. F. and Cussler, E. L.,** Selective membrane transport, *AIChE J.,* 19, 736, 1973.

52. **Wong, K. H., Yagi, K., and Smid, J.,** Ion transport through liquid membrane facilitated by crown ethers and their polymers, *J. Membr. Biol.,* 18, 379, 1974.

53a. **Sadakane, A., Iwachido, T., and Toei, K.,** Distribution studies of alkali metal picrates, *Bull. Chem. Soc. Jpn.,* 45, 432, 1972.

53b. **Sadakane, A., Iwachido, T., and Toei, K.,** *Bull. Chem. Soc. Jpn.,* 51, 629, 1978.

54. **Peppard, D. F., Gray, P. R., and Mason, G. W.,** The solvent extraction behavior of the transition elements. I. Order and degree of fraction of the trivalent rare earths, *J. Phys. Chem.,* 57, 294, 1953.

55. **Peppard, D. F., Mason, G. W., Maier, J. L., and Driscoll, W. J.,** Fractional extraction of the lanthanides as their dialkyl ortho-phosphates, *J. Inorg. Nucl. Chem.,* 4, 334, 1957.

56. **Peppard, D. F., Farraro, J. R., and Mason, G. W.,** Hydrogen bonding in organophosphoric acids, *J. Inorg. Nucl. Chem.,* 7, 231, 1958.

57. **Peppard, D. F.,** *The Rare Earths,* Spedding, F. H. and Daane, A. H., Eds., John Wiley and Sons, New York, 1961, 38.

58. **Topp, N. E.,** *Chemistry of the Rare-earth Elements,* Elsevier, Amsterdam, 1965, 31.

59. **Goto, T. and Smutz, M.,** Separation factors for solvent extraction processes. The system of $1M$ di(2-ethyhexyl) phosphoric acid (in AMSCO 125-82)-Pr-Nd salts as an example, *J. Inorg. Nucl. Chem.,* 27, 1369, 1965.

60. **Nair, S. G. K. and Smutz, M.,** Recovery of lanthanum from didymium chloride with di(2-ethyl-hexyl)phosphoric acid as solvent, *J. Inorg. Nucl. Chem.,* 29, 1787, 1967.

61. **Lenz, T. G. and Smutz, M.,** The extraction of neodymium and samarium by di(2-ethylhexyl) phosphoric acid from aqueous chlorides, perchlorate and nitrate systems, *J. Inorg. Nucl. Chem.,* 30, 621, 1968.

62. **Shiokawa, J., Matsumoto, A., Takatsuji, K., Morito, S., and Hirashima, Y.,** Extraction of europium by di(2-ethylhexyl) phosphoric acid, *Kogyo Kagaku Zasshi,* 74, 14, 1971.

63. **Pierce, T. B., Peck, P. F., and Hobbs, R. S.,** The separation of the rare earths by partition chromatography with reversed phases. Part II. Behavior of individual elements on HDEHP-corvic columns, *J. Chromatogr.,* 12, 81, 1963.

64. **Rahn, R. W. and Smutz, M.,** Development of laboratory scale continuous multistage extractor, *Ind. Eng. Chem., Proc. Des. Dev.,* 8, 289, 1969.

65a. **Jiang, C., Yu, J., and Zhu, Y.,** Mass transfer mechanism of extraction with liquid membranes, *Chem. Abst.,* 98, 98:111303u, 1983.

65b. **Yu, J., Wang, S., Jiang, C., and Zhu, Y.,** Advancing front model of liquid surfactant membranes, *Chem. Abst.,* 101, 101:154033d, 1984.

66. **Teramoto, M., Sakuramoto, T., Koyama, T., Matsuyama, H., and Miyake, Y.,** Extraction of lanthanoids by liquid surfactant membranes, *Sep. Sci. Technol.,* 21, 229, 1986.

67a. **Araki, T., Okazawa, T., Kubo, Y., Ando, H., and Asai, H.,** Separation of lighter rare earth metal ions by centrifugal countercurrent type chromatography with di(2-ethylhexyl) phosphoric acid, *J. Liq. Chromatogr.,* 11, 267, 1988.

67b. **Araki, T., Okazawa, T., Kubo, Y., Ando, H., and Asai, H.,** Separation of lighter rare earth metal ions by centrifugal countercurrent type chromatography with D2EHPA as stationary phase, Proc. Chem. Soc. Jpn., Chugoku-Shikoky, Div., Nov. 22, 1987, 46.

68. **Araki, T., Okazawa, T., Kubo, Y., Asai, H., and Ando, H.,** Further results on behaviors of rare earth metal ions in centrifugal partition chromatography with di(2-ethylhexyl) phosphoric acid, *J. Liq. Chromatogr.,* 11, 2473, 1988.

69. **Araki, T., Okazawa, T., and Kubo, Y.,** CPC and liquid-membrane transport of rare earth metal ions — possibility for purification and sensing, Proc. 2nd CPC Colloq. (Progress and applications of CPC), Kyoto, June 4, 1988, 67.

70. **Akiba, K., Sawai, S., Nakamura, S., and Murayama, W.,** Mutual separation of lanthanoid elements by centrifugal partition chromatography, *J. Liq. Chromatogr.,* 11, 2517, 1988.

71. **Araki, T., Okazawa, T., Asai, H., and Kubo, Y.,** Correlation between liquid membrane transport and CPC separation — behaviors of heavier rare earth metal ions, Proc. Chem. Soc. Jpn., Chugoku-Shikoku-Kyushu Div., Oct. 10, 1988, 54.

72a. **Li, N. N.,** Separation of hydrocarbons by liquid membrane permeation, Membrane Processes Ind. Biomed., Proc. Symp., 1971, 175.

72b. **Babcock, W. C., Baker, R. W., Kelly, D. J., and LaChapella,** Coupled transport membranes for metal recovery, phase II, *Proc. ISEC'80,* 80, 1980.

72c. **Chiarizia, R. and Danesi, P. R.,** A double liquid membrane system for the removal of actinides and lanthanides from acidic nuclear wastes, *Sep. Sci. Technol.,* 22, 641, 1987.

72d. **Nakano, M., Takahashi, K., and Takeuchi, H.,** A method for continuous operation of supported liquid membrane, *J. Chem. Eng. Jpn.,* 20, 326, 1987; Takahashi, K., Nakano, M., and Takeuschi, H., Chromium(VI) recovery from sulfuric acid solution with supported liquid membrane, *Kagaku Kougaku Ronbunshu,* 13, 657, 1987.

72e. **Kataoka, T., Nishiki, T., Yamaguchi, M., and Zhong, Y.,** A simulation for liquid surfactant membrane permeation in a continuous countercurrent column, *J. Chem. Eng. Jpn.,* 20, 410, 1987.

72f. **Dworzak, W. R. and Naser, A. J.,** Pilot-scale evaluation of supported liquid membrane extraction, *Sep. Sci. Technol.,* 22, 677, 1987.

72g. **Teramoto, M., Matsuyama, H., Takaya, H., and Asano, S.,** Development of spiral-type supported liquid membrane modules for separation and concentration of metal ions, *Sep. Sci. Technol.,* 22, 2175, 1987.

# INDEX

# INDEX

## A

Acetylcholine, 42
Active transport, 12, 78, 87—98
  in biomembranes, 12—14
  defined, 14—15, 89
  energy conversion in, 89—91
  irreversible thermodynamics of, 87—89
  mechanism of, 18—21
  mechanistic principles of, 87
  models of, 91—98
  primary, 15—17
  secondary, 17—18
Acyclic crown compounds, 63—64, see also specific
    types
Acyclic crown ether, 68
Acyclic host-guest chemistry, 4
Acyclic podands, 55
Acyclic polyether antibiotic, 52
Adenosine, 113
Adenosine diphosphate (ADP), 16, 44, 108
Adenosine monophosphate (AMP), 44
Adenosine triphosphatase (ATPase), 10, 15, 17, 20
Adenosine triphosphate (ATP), 16
  hydrolysis of, 15, 23
  macrocyclic polyamines and, 108
  phosphorylation by, 20
  selectivity for, 108
  sensitivity to, 108
  synthesis of, 8, 10
ADP, see Adenosine diphosphate
Alanine, 56
Alcohols, 144—145, see also specific types
Alkali cations, 53, see also specific types
  earth metal, 53, 134
  separation of, 127—129
Alkali metal complex, 68
Alkali metal ions, 4, see also specific types
Alkali metal picrates, 204
Alkylammonium, 108
Alkyl chains of lipids, 10—11
Alkylphenyl boric acid, 114
Ambient pH, 178
Amines, 44, see also specific types
Amino acids, 42, 64, see also specific types
  free, 113
  salts of, 70
  symport of, 113
Ammonium cations, 42, 56, 64
Ammonium salts, 106
AMP, see Adenosine monophosphate
Amphiphiles, 178, 180, see also specific types
Amphiphilic crown ethers, 172
Anion cryptate, 66
Anion exchange, 66
Anionic surfactant, 143, 144
Anions, see also specific types
  extractants for, 55

  lipophilic, 66
  organic, see Organic anions
  transport of, 32, 44—45, 53, 64—68
Anion-selective electrodes, 106
Anion sensors, 106
Antibiotics, 3, 52, 104, 106, see also specific types
Anticancer drugs, 4
Antigen-antibody interactions, 158
Antiport, 15, 22, 31, 70, 92
Applications, 4, 37—47, 137—138, see also specific
    types
  in chemistry, 4
  in electron transport, 46
  in gas separation, 45—46
  in ion-selective electrodes, 47
  in medicine, 4
  in metal ion recovery, 39—42
  in neutral molecule transport, 46—47
  in organic anion transport, 44—45
  in organic cation transport, 42—44
Arabinose, 114
Armed host molecules, 55
Armed macrocycles, 60—63
Aromatic compounds, 116, see also specific types
Arrhenius plots, 184
ATP, see Adenosine triphosphate
ATPase, see Adenosine triphosphatase
Azacrowns, 68, 108
Aza-macrocycles, 61
Azobenzene-containing amphiphiles, 180
Azobis (benzocrown ethers) (butterfly crown ethers),
    170

## B

Back-pack (facilitated) transport, 2, 38, 83—87
Bacteriohodopsin, 17
Barium, 60, 62
Barrier forming, 8
Bathocuproine, 162
Bathophenanthroline, 68
Belousov-Zhabotinskii reaction, 142
Benzene, 114
Benzene-dicarboxylic acid, 66
Benzylammonium, 108
Bicyclic cryptands, 55
Bilayer-intercalated clay films, 183—188
Bilayers, 4, 9, 146, 178
Biogenetic amines, 44, see also specific types
Biological carrier proteins, 56
Biomembranes, 8
  asymmetry of, 10
  fluid nature of, 8—12
  structure of, 8—12
  transport in, see under Transport
Biominetic sensory systems, 33
Biotechnology, 4
Bovine serum albumin, 116

# RETURN TO → CHEMISTRY LIBRARY
100 Hildebrand Hall          642-3753

| LOAN PERIOD 1 | 2 | 3 |
|---|---|---|
| 7 DAYS | 1 MONTH | |
| 4 | 5 | 6 |
| | | |

## ALL BOOKS MAY BE RECALLED AFTER 7 DAYS
Renewable by telephone

## DUE AS STAMPED BELOW

UNIVERSITY OF CALIFORNIA, BERKELEY
FORM NO. DD5, 3m, 12/80     BERKELEY, CA 94720
®s